TRIAL BY FIRE: WORLD WAR II AND THE FOUNDERS OF MODERN NEUROSCIENCE AND PSYCHOPHARMACOLOGY

Wallace B. Mendelson

Pythagoras Press

New York

About the Author

Wallace Mendelson, MD is Professor of Psychiatry and Clinical Pharmacology (ret) at the University of Chicago. He is a Distinguished Fellow of the American Psychiatric Association, and a member of the American Academy of Neuropsychopharmacology. He has been director of the Section on Sleep Studies at the National Institute of Mental Health, the Sleep Disorders Center at the Cleveland Clinic Foundation, and the Sleep Research Laboratory at the University of Chicago. He is the author of eleven books and numerous professional papers. He is currently a Fellow of the Faculty of History and Philosophy of Medicine and Pharmacy at the Worshipful Society of Apothecaries, London. Among his honors have been the Academic Achievement Award from the American Sleep Disorders Association in 1999 and a special award for excellence in sleep and psychiatry from the National Sleep Foundation in 2010. More information about Dr. Mendelson and his work is available in Wikipedia at https://en.wikipedia.org/wiki/Wallace B. Mendelson and on his website at http://zhibit.org/wallacemendelson.

Disclaimer and Conflict of Interest Statement

CONFLICT OF INTEREST: Dr. Mendelson has no financial arrangements with any pharmaceutical company marketing any medicines mentioned in this book.

DISCLAIMER: This book contains information on a variety of psychiatric and neurologic disorders as well as their treatments. It is not a substitute for medical evaluation and treatment. Any person who believes that they may have any of the disorders mentioned in this book should consult their physician.

CONTENTS

PREFACE

In *The Curious History of Medicines in Psychiatry* (Mendelson, 2020a), which deals with the birth of modern psychopharmacology in the 1950s, I briefly described the lives of some of the key figures involved, and was struck with how many had been caught up in World War II in one way or another. Some like Henri Laborit, who later initiated the use of chlorpromazine for psychoses, and John Cade, who made the modern discovery of lithium for bipolar disorder, had been military physicians. Laborit had been aboard a French destroyer struck by torpedoes during the evacuation of Dunkirk and was awarded a Military Cross for his actions during the sinking. Cade had volunteered for service in the Australian Army, and served in Singapore until he was captured and spent several years as a prisoner of the Japanese. In both cases, one could see the influence of their wartime experiences on their later work: after the development of chlorpromazine, Laborit had quickly suggested that it might be useful in managing stress in wounded soldiers on the battlefield (an unsuccessful, even disastrous, endeavor). Cade, while caring for his fellow prisoners of war, had noticed their fluctuating levels of mental clarity, and wondered if this might be related to changing amounts of circulating substances, foreshadowing his later studies which led to lithium.

Others crucial to the development of psychopharmacology were affected by the war in various ways: Frank Berger, who went on to discover meprobamate, the first modern tranquilizer, was a young physician in the Czechoslovak National Institute of Health who fled the Nazi occupation, arriving in Britain in 1939, where he first worked providing medical care in refugee camps. Leo Sternbach, born in what is now Croatia, was an organic chemist who had experienced antisemitism firsthand in Poland. He was working for the Hoffman LaRoche Co. in Basel when Germany invaded Greece and Yugoslavia in 1941. Facing an increasingly precarious situation, he fled through France and Portugal and came to America, where he went on to discover the first benzodiazepine tranquilizer. The chemist Charles Suckling, who worked in Britain's secret atomic bomb research and served in the Home Guard when away from the laboratory, later applied some of his new-found techniques to develop the anesthetic halothane.

Still others had had experiences in earlier wars. William A. Hammond, who made observations with lithium in bipolar disorder in the 1870s, had been Surgeon General of the Union Army during the American Civil War (Mendelson, 2020b). Hans Berger, serving in the German cavalry in 1892, was almost killed in a riding accident. His resulting curiosity about the workings of the nervous system led to his discovery of the human EEG, and his story continued to include suffering at the hands of the Nazis who took over his university in the 1930s. Constantin von Economo, the Austrian psychiatrist and neurologist who described encephalitis lethargica and the role of hypothalamic structures in sleep and waking, had been an army pilot in northern Italy in 1916 (Mendelson, 2017). Charles Sherrington, whose studies with Edgar Adrian clarified the processes of nerve cell communication, worked in a shell factory 13 hours a day during World War I, while Adrian took care of war casualties at St. Bartholomew's Hospital. Alexander Fleming, whose later discovery of penicillin profoundly affected psychiatry by providing a treatment for tertiary syphilis, served in World War I where his studies of evolving wounds influenced his thinking leading to his later discoveries (Mendelson, 2020b).

In summary, I was struck by the wartime background of many of the foundational figures in psychopharmacology (and in the case of Berger, electrophysiology). Since the broader field of neuroscience also developed in the post-World War II years, I wondered if the same might be true of some its founders as well. I was already aware that Otto Loewi, whose experiments confirmed chemical transmission across synapses, had been imprisoned by the Gestapo on the night of the German invasion of Austria, was stripped of all his possessions, and later allowed to emigrate on the condition that he transfer the funds from his Nobel Prize to a Nazi-controlled bank. Similarly familiar was the story of Alfred Loomis the American self-taught scientist, who discovered the K-complex in the EEG and first classified the sleep stages, and then went on to play a crucial role in developing military radar in World War II. On further exploration, many others appeared, and several seemed to return to their memories of war in older age. Alan Hodgkin's experiences living in wartime Britain and engaging in radar research had impacted him in such a way that much later at age 78 he felt the need to come back to it in his last major work, *Chance and Design: Reminiscences of Science in Peace and War.* Arvid Carlsson as a medical student examined concentration camp prisoners who had been released to neutral Sweden; the experience so impressed him that he talked about it at length in his Nobel Prize biography 56 years later. Starting in his 60s, Henri Laborit wrote extensively on the applications of neuroscience to aggression and violence. On further exploration it became clear that indeed many of the crucial figures in neuroscience as well as psychopharmacology had been affected by—and in turn influenced—the war in a number of ways. This book is the story of their lives.

Collections of short biographies have been used over the years for expressing a belief system or for characterizing an era. The classic example is the Catholic *Lives of the Saints*, which emerged in the seventeenth century, and continues to appear in modern forms (Butler, 2015). Collective biographies have been written on diverse subjects ranging from world-class leaders (Shoup, 2005) to medieval Jewish physicians in the Muslim world (Lev, 2021). In a recent example, *The Black Bottom Saints* (Randall, 2020), a combination of fact and imaginative biographies of 52 characters was used as a way of characterizing and memorializing Detroit's historic Black Bottom Neighborhood.

In this book, not all the characters were saintly (as they were not in Black Bottom), and every effort has been made to be factual. Nonetheless it follows a tradition that a collection of short life histories can be an effective way of capturing the sense of an era. In this case the goal is to describe how World War II and its surrounding years affected the personal and professional lives of many of the people who founded two new interrelated fields. The remarkable thing is the variability of their experiences. As we will see in Chapter Two, sometimes the war represented more of a hiatus in ongoing research which was resumed afterwards. In other cases, it led to displacement or imprisonment. Sometimes the result was devastating and culminated in suicide, while in others it was a stimulus leading to advances in research. Often these experiences are known only to specialists these fields. The hope here is that by collecting and organizing them here in this way, these stories will not be lost, and will give a better understanding of the lives of the people who in a very real sense changed the world of science and medicine.

There are many sources on the history of neuroscience (Glickstein, 2014) and psychopharmacology (Braslow and Marder, 2019) describing the crucial experiments and findings. The first section of the Appendix of this book provides a short summary of the major events in chronological order, to give a sense of structure for persons to whom it may be less familiar. The intent of this book is to be complementary, focusing on the World War II era and providing narratives which give insight into the lives of the discoverers who were particularly affected by the war, in a compact, non-technical manner. As an aside, there has been interest in the process of narrative as being an inherently human method for understanding the world, with a counterpart in the functional anatomy in the nervous system; some neuroscientists have suggested that for this reason it might be a particularly good method for communicating between scientists and society (see Appendix, section 2).

It should also be mentioned that while writing this book it quickly became apparent that commenting on the wartime experiences of neuroscientists and psychopharmacologists brings up a variety of related topics. There are histories, for instance, of psychotropic drugs being used to enhance the performance of combatants, or of being used as weapons, or even the broader question as to the degree to which the post-war blossoming of neuroscience was significantly funded by the military. These are beyond the scope of this book, but in order to increase awareness of the many related topics, they are briefly summarized in the final chapter, along with references for those interested in pursuing them.

Finally, a historical note: It has sometimes been said that the modern world began in the decades around 1500, during which the New World was discovered, Martin Luther started the Reformation, and the fall of Constantinople marked the end of the Byzantine Empire. To this list might be added a fourth—the life of Paracelsus (1493-1541), a warrior-physician who played a crucial role in what became known as the Medical Renaissance (Mendelson, 2020c). The son of a Swiss chemist and physician, he obtained his medical degree from the University of Ferrara in about 1515, and shortly thereafter volunteered as a military surgeon. Over the next seven years he served in wars involving Holland, Venice and Denmark. At one point he was taken prisoner by the Tartars in Russia, then got away from his captors in Lithuania and moved on to Hungary. Later he joined the Italian army, and ultimately traveled through the Middle East and the Holy Land. His travels exposed him to many different cultures, and led him to a firm belief that much was to be learned from the practical experience of barbers, gypsies, sorcerers and even medicine men of wandering tribes. On his return he briefly taught at the University of Basel, where it became clear that he did not fit in well with scholastic life, and left town hastily in 1528 after having defied the authorities (and perhaps not coincidentally, shortly after the death of a prominent patient). He resumed his life as an itinerant physician, learning and teaching along the way. He was among the first to discard the notion of panaceas, magical medicines which could cure all illnesses, and focus instead on drugs for specific disorders. Similarly, he recognized the importance of dose, emphasizing that the same medicine might be helpful in one amount and toxic in another. He recommended mercury as a treatment for syphilis, iron compounds for 'poor blood', and antimony as a purgative. In doing so, he was among the first to bring together medicine and chemistry in a meaningful way.

The tradition of the Medical Renaissance continued with Leonardo da Vinci (1452-1519), whose anatomical drawings were informed by his understanding of how the eye sends visual information to the brain, the French physician Ambrose Paré (1510-1590), who improved the treatment of wounds and surgical practice, Vesalius (1514-1564), who created detailed anatomic studies deriving from dissection, and many others. Early pharmacopoeias, often deriving from Paracelsus's work, came out from centers of learning in Nuremburg (1546) and Basel (1561). And much of this began with information gathered by a military surgeon in his wanderings around most of Europe and the Middle East.

Figure Pre-1: Paracelsus performing head surgery.

PROLOGUE: SIGMUND FREUD

At first glance the story of Sigmund Freud (1856-1939) may seem out of place in a history of the leaders of neuroscience and psychopharmacology in the World War II era. Freud was of an earlier generation—to put the timing in perspective, he was born before the American Civil War. The later dominance of the psychoanalytic school of thought which he founded was if anything, a hindrance to the acceptance of new pharmacologic treatments in the mid-twentieth century. When imipramine, the first tricyclic antidepressant became available, for instance, analytically-oriented psychiatrists were critical—since depression is the result of a loss, they argued, how could a medicine replace a loss? (See the companion book *The Curious History of Medicines in Psychiatry;* Mendelson, 2020a.) Nonetheless, Freud's ongoing development of psychoanalysis was profoundly affected by World War I, and serves as a model of how living in the proximity of war can influence later thinking. Secondly, like Otto Loewi (Chapter One), he was terrorized by the Gestapo within a day of the Nazi takeover of Austria in 1938, and his response to it was a remarkable outburst of productivity. For these reasons, his experiences seem a fitting background for the story of the next generations, which are the focus of this book.

At the outbreak of World War I, the 58-year-old Sigmund Freud appeared to be at the peak of his career, at the head of the growing psychoanalytic movement. Initially, like many in Austria-Hungary, he was enthusiastic, but as his sons were drafted into the Hapsburg army and he began to recognize the wholesale destruction that was taking place, he became weary and discouraged, and expressed the widespread disillusionment in his book *Thoughts for the Times on War and Death* (Freud, 1916). Especially hard was recognizing the breakdown of the *Pax Britannica*, which had upheld Western values and provided relative stability for decades (Freud Library, 1985).

Freud's disillusion with the war affected his thinking in at least two important ways. Earlier in his career he had been a lecturer in neuropathology and had published on cocaine and on clinical conditions such as aphasia. As he moved into studies of the mind, his 1895 *Project for a Scientific Psychology* made clear his commitment to determine the physiologic basis of psychological processes (Gay, 1995). His topographical view of the mind, as formulated in the years leading up to World War I, retained some biological underpinnings consistent with this history. Now, rather than focusing primarily on the role of the sexual impulse and its repression in the origin of neuroses, he came to believe that much of human behavior could be explained by the opposition of two powerful forces: 'eros', the drive for love, and 'thanatos', the predilection for aggression and violence. Secondly, he began to move from a viewpoint centering on the individual and the relationship to parents, to a new one involving the role of society in shaping morals and helping keep aggression under control (for a full description, please see Appendix 3). He also participated in the post-war review of the apparently harsh manner in which wartime neuroses had been handled, following accusations that Julius Wagner-Jauregg, an eminent psychiatrist and future Nazi supporter, had seemingly inflicted suffering on his patients. Freud began to emphasize the importance of the community in providing treatment for mental illness, and as time went on the second generation of psychoanalysts began to provide free community-based clinics (Danto, 2016). In 1930 he wrote *Civilization and its Discontents*, in which he dwelt on the drive of individuals toward impulsive behavior including violence, and the sometimes unsuccessful role of society in curbing these impulses. Phrased differently, he began to take the view that human unhappiness results not only because of repressed sexuality, but also because of the restraints society places on aggressive and destructive instincts (Shaw, 2021).

Freud was struggling with these concepts when once again he began to experience the pathway to war, with the growing political influence and violence of the Nazis. In 1932, Albert Einstein wrote him from Berlin, asking if he thought war was so inherent in human nature that it is inevitable, or whether it can be prevented. In a famous reply, Freud emphasized both the depth to which we have instincts for violence, but also the hope for human culture which can 'master our instinctive life'. In Freud's words, 'He (Einstein) understands as much about psychology as I (Freud) do about physics, so we had a very pleasant talk' (Ninivaggi, 2012). The two followed very different paths after that; shortly after Hitler became Chancellor of Germany in 1933, Einstein fled for England and later the U.S., while Freud chose to stay in Vienna. He had been diagnosed with cancer of the jaw a decade earlier, and was determined to die in Vienna, where he would 'disappear from this world with decency'. By that time Freud was experiencing the growing persecution firsthand. The book burnings which took place beginning in May 1933 were devoted to 'un-German' works, of which Freud's were so high on the list that a special 'fire-oath' denouncing 'overvaluation of sexual activity' was recited as the books were thrown into the flames.

The day after the *Anschluss,* the Nazi takeover of Austria in March 1938, the 82-year-old Freud's home was raided by the Gestapo, who took his money, valuables and passport. Left with little but his sense of humor, he commented at the time 'I never received as much for a house call' (U.S. Holocaust Museum, 2021). Freud's sons had already left, but his daughter Anna was picked up by the Gestapo, taking with her a bottle containing a lethal dose of Veronal (a barbiturate) in case she were to be tortured (N. Tucker, 2019). She was released later in that day after protests by two ex-patients, William Bullitt, the American ambassador to France, and Princess Marie Bonaparte, great-granddaughter of Napoleon, as well as a phone call from the American consul in Vienna to the Gestapo during her interrogation.

Figure P-1: *Sigmund and Anna Freud, on vacation in the Italian Dolomites in 1913. This was a happy period in Freud's life, while he was experiencing success and before the disillusion that came with World War I. Some years later, after the Anschluss in 1938, Anna's arrest and later release by the Gestapo led to the decision to leave Vienna for London.*

Freud was now committed to exile, but the problem was obtaining permission. Fortunately, help came from several sources. One was once again his princess-patient, who paid the Nazis a considerable amount of money to allow him to go (Latson, 2015). He is also said to have obtained help from an unlikely source, Anton Sauerwald, a German official who had been appointed to oversee Freud's possessions and publishing business. Sauerwald was a 40-year-old chemist whose professor Josef Herzig had been friendly with Freud, and he himself read some of Freud's works and came to respect him. He sold some of Freud's possessions to raise money, which he used to pay for favorable action on visas for him and 16 members of his family (Cohen, 2012). The story is a complicated one—he was a Nazi whose hobby was bomb-making, and who after the war was charged with war crimes including stealing the family's assets. A letter from Anna Freud in his support ultimately aided in his release from jail.

On June 4, 1938, Freud was allowed to take the Orient Express to Paris, as a first stop on the way to London. Before he could board, the Gestapo required him to sign a document freeing them from any blame for his treatment. They perhaps did not appreciate his sense of humor when he wrote 'I can heartily recommend the Gestapo to anyone' (Cohen, 2012).

Once in London, Freud continued to be troubled by cancer, and Sauerwald, who had come to visit him, arranged for his Vienna doctor to travel there to perform an operation which Freud thought extended his life by a year. He was distressed by not knowing the fate of four of his sisters, for whom Sauerwald had been unable to obtain visas. It was only several years later that it was learned that all had perished in the camps.

In his remaining year Freud's work was very productive. He saw patients in a room in his new Hampstead home which was designed to resemble his old office in Vienna, complete with the original couch. During this time, he received visits from luminaries such as H.G. Wells, Salvador Dali and Virginia Woolf. He completed his long-unfinished book *Moses and Monotheism*, and a synopsis of his work, *An Outline of Psychoanalysis*, which came out posthumously. When asked how had managed to do so much in the past year, his reply was 'Thank the Fuhrer' (Cohen, 2012). He died three weeks after war was declared in September 1939.

Figure P-2: *Freud's couch, on which patients would recline while undergoing psychoanalysis. It became something of a symbol of his work, and when Freud left Vienna, his antiquities were sold to raise money to allow him to take it and some of his books to London. They now reside in Hampstead in the Freud Museum.*

CHAPTER ONE: THE PRE-WAR YEARS

This book is organized in a generally chronological format, according to when the major discoveries took place. In this section we will look at discoverers whose main work in neuroscience was in the years immediately preceding the war, and whose lives were often significantly affected by the rise of Nazi power.

Otto Loewi (1873-1961) and the stuff that dreams are made of

Otto Loewi was born in Frankfort, Germany, the son of a well-to-do wine merchant. He was drawn to classical languages and art, and planned to study art history, but was persuaded by his father to go to medical school. Once at the University of Strasbourg in 1891, he tended to skip his required classes in favor of those in the humanities. He barely squeezed through the first major examination, and had to spend a remedial year before gaining enthusiasm for his studies.

After graduation in 1896, Loewi worked at the City Hospital in Frankfurt, where like Paul Ehrlich before him in Berlin (Mendelson, 2020b), he became very impressed with the devastation caused by tuberculosis, for which there was no effective treatment. He decided to go into research, first at the University of Marburg where he studied glucose and protein metabolism for six years, and then in

1909 had risen to become the Chair of Pharmacology at the University of Graz in Austria.

In 1902 he visited England, spending time at University College, London, where he met two colleagues who greatly influenced his life. One was Henry Dale, who was interested in the sympathetic and parasympathetic nervous systems, components of the autonomic nervous system which operates largely unconsciously and regulates respiration, heart rate, digestion, sexual arousal and other bodily processes. He later identified the neurotransmitter acetylcholine, recognizing that it produced many of the effects of stimulating the parasympathetic system. Loewi also met the Cambridge medical student Thomas Renton Elliott, who observed that epinephrine produced effects associated with the sympathetic nervous system, and thought that it might be released by some of its nerve cells.

Returning to Graz, Loewi brought back with him a newfound interest in autonomic function. In the following years he studied the effects of drugs on the role of the vagus nerve (part of the parasympathetic nervous system) in regulating heart function, using a model in which frogs' hearts, which continued to beat after extraction, were studied in saline baths in glassware. He eventually became involved in an ongoing debate among scientists about whether nerve cells communicate with each other and with other tissues by means of releasing chemicals or by electrical impulses. Each view had its advocates: it was known that chemicals could produce physiological effects such as contracting muscles, for instance, though no physiological release of substances had been identified. Similarly, it had been found that electrical stimulation could produce physiological reactions.

By 1920 Loewi was struggling to reconcile the views of what came to be known as 'wet' and 'dry' neurophysiology (less elegantly referred to as 'soup' and 'sparks'), when an idea came to him in his sleep. Drifting off while reading (much as August Kekulé had the night he discovered the structure of the benzene ring; Mendelson, 2020b), Loewi awakened with an idea for an experiment. He hastily scribbled down his ideas, and returned to sleep. The next morning to his consternation he was unable to read his hastily written notes. After spending what he later described as the longest day of his life, he went to sleep the following night, and later awakened with the same idea. This time he arose, went to his laboratory, and conducted the experiment which changed neurophysiology and won him the Nobel Prize.

Loewi's experiment, in which he combined a hypothesis he had been thinking about for some time with a physiologic preparation he had already been working with, had the elegance of simplicity. He made two physiological preparations of frogs' hearts. In one, the vagus nerve was intact and connected to the heart; in the other the heart was completely isolated. He electrically stimulated the vagus nerve going to the first heart, whose rate of beating consequently slowed down. He then took fluid from the chamber in which it was immersed, and added it to the bath containing the second heart, which itself then slowed down. It appeared that some substance released by the vagus nerve had altered the heart's rate of beating. Loewi called this chemical '*Vagusstoff*' (Vagus substance), which he later found to be the neurotransmitter acetylcholine. As a result of these studies Loewi as well as Dale, whose work complemented Loewi's, were awarded the Nobel Prize in 1936.

March 12 of 1938 brought the Nazis *Anschluss,* the invasion and annexation of Austria, and that night Loewi was awakened by a dozen storm troopers; he and two of his sons were arrested and imprisoned by the Gestapo. (See the Prologue for Sigmund Freud's experience with the Gestapo the day after the *Anschluss*.) In addition to suffering this fate as a Jew, Loewi had also violated a ban on citizens accepting the Nobel Prize, promulgated by Hitler as a result of his experience with Carl von Ossietzky, who won the Nobel Peace Prize in 1935 for exposing Germany's illicit rearmament (Chapter Four). News of Loewi's arrest reached the International Physics Congress in Zurich, and the ensuing protests ultimately led to his release two months later, but only after he was forced to turn over his Nobel Prize money to a Nazi bank. He left for England in September of 1938, but his wife was detained in Austria in an effort to get the rights to her property in Italy, and was not able to leave the country until 1941.

In 1940 Loewi emigrated to the U.S., where he became a citizen, taught at New York University and was an honored member of the Woods Hole, Massachusetts scientific community. He passed away on Christmas Day of 1961.

Figure 1-1: Otto Loewi's 1936 Nobel Prize in Physiology or Medicine, shared with Henry Dale.

In closing this section on Otto Loewi, it should be mentioned that his discovery later generated a number of colorful stories. Some have emphasized the role of chance, arguing that the time of night, the season, and even the particular species of frog were crucial to his success; under other circumstances the acetylcholine might have been chemically broken down more quickly and not affected the recipient frog's heart. On one occasion Henry Dale suggested that on the second night that Loewi had his famous dream, he merely got up and wrote down more detailed and legible notes, rather than going to the laboratory and performing a nocturnal experiment. It is also not entirely clear whether the experiment came to him in a dream *per se*, or whether he 'merely' awakened from sleep with the idea. Perhaps we will never know, but the result was that a man who was once an unpromising medical student initiated a revolution in our understanding of how nerve cells communicate, sustainable even in the face of Nazi repression, which has influenced the field of neuroscience to this day.

Hans Berger and the pursuit of 'psychic energy'

Hans Berger, born in 1873 in Neuses, Germany, was the son of the head physician of a nearby asylum, and the grandson of the poet Friedrich Rückert, to whose work he became very attached. He was a shy, introspective youngster with little interest in medicine, but was drawn to astronomy. After a semester at the university in Jena in 1892, for reasons which have never been well described, he interrupted his studies and enlisted in the cavalry. There he had an experience which changed his life. One day his horse unexpectedly reared up, throwing him to the ground in front of a horse-drawn cannon. Though it appeared likely that he would be run over, the driver managed to stop in time, and Berger was saved. At apparently that same time, his sister, who lived some distance away, was overcome with a feeling that her brother was in danger, and appealed to her father to send him a telegram asking if he were okay. The telegram seemed to Berger to be evidence that 'spontaneous telepathy' had taken place, and after returning to Jena he changed his studies to medicine with a goal of determining what kind of 'psychic energy' might have been involved. Following graduation in 1897, he worked in the psychiatry and neurology clinic in Jena; after eight years devoted to studying blood flow in the brains of patients with skull fractures, he turned his attention to whether electrical activity could be measured in the human brain, as had previously been found in animals.

Berger's studies were interrupted by World War I, in which he worked as an army psychiatrist; afterwards he returned to Jena and ultimately became the chief of psychiatry there. Some thought that he appeared to lead a kind of double life: outwardly, he seemed rigid and humorless, while his private time was filled with poetry and recording spiritual thoughts in his diary. He was also very persistent, and in 1924 he used a vacuum tube amplifier to record the first electroencephalogram, a term he coined, in a 17-year-old boy who was undergoing neurosurgery. He refined the technique, later recording from the periosteum membrane covering the skull, and then from the surface of the scalp, where he studied, among others, his own children. By 1929 he had gained enough experience that he felt comfortable publishing his first paper, and over the next years described EEG changes with age, mental activity and sleep, and in persons with brain tumors. Initially he was met with indifference and skepticism. By the mid-1930s he gained the support of the English electrophysiologist Edgar Adrian (Appendix 1) and gave talks with him at meetings. He hoped to travel with him to the U.S., but was unable to because of the imminence of war. By that time the EEG was recognized internationally as a useful clinical tool.

In 1938 the Nazi control of the university became the dominant element in his life. Exactly what happened is a matter of debate. Early accounts indicated that at age 65 he was forced to retire, and told that he could never again work on the EEG. Later records suggested that he remained at the university, and was even part of the committee that chose his successor, a prominent Nazi named Berthold Kihn, who was later accused of being involved in 'euthanasia' deaths. Berger then served on what was known as the Court for Genetic Health, supervising sterilization of psychiatric patients, a post he outwardly appeared to accept with enthusiasm. His diaries were later found to include anti-Semitic thoughts. He also contributed financially to the SS. Whether he was truly the victim as he was first portrayed, was pressured for survival into cooperation, or was even a willing participant is still debated (Zeidman, 2014). It is known that he became very depressed and suffered from a troubling skin disease, and in 1941 he hung himself in his own clinic.

Alfred Lee Loomis and the classification of the sleep stages

Alfred Loomis (1887-1975) came from quite a different background than Loewi or Berger. Born in Manhattan to a socially prominent family, many of whom were physicians, he went on to study mathematics at Yale before going to Harvard Law. After graduating in 1912 he practiced corporate law until joining the Army when the U.S. entered World War I in 1917, and was assigned to the Aberdeen Proving Grounds in Maryland. During his time there he invented a device which became known as the Aberdeen Chronograph, which measured the velocity of shells by firing them through revolving discs of aluminum covered with paper. He also became friendly with Robert W. Wood, a physicist from Johns Hopkins with whom he would later pursue his interests in science.

After the war Loomis and his brother-in-law Landon Thorne acquired a struggling investment bank, which grew to become the leader in dealing with electrical utilities, managing up to 15 percent of all American securities. Sensing that the Wall Street Crash of 1929 was coming, they sold their investments for gold, and after the crash made even more money by purchasing stocks which were at an all-time low. In the years that followed, Loomis prospered financially, but he began to miss the excitement of invention he had felt at Aberdeen. His interests returned to science and his fascination with building new devices. He had little formal scientific education, but in the tradition of the English gentleman-scientist he built a laboratory, equipped better than those at most universities, near his mansion high on a hill in the exclusive community of Tuxedo Park, New York. Just as Hans Berger in some senses led a double life—the austere professor by day, while a writer of rich spiritual thoughts by night—Loomis was the canny, successful businessman by day, but the passionate amateur scientist in the evening. As time went on, his laboratory became well known, and at various times he was visited by a remarkable group of luminaries including Albert Einstein, Niels Bohr and Enrico Fermi. In those years he pursued many interests, including quartz-crystal chronometers, spectrometry of chemicals such as formaldehyde, and high-frequency sound waves.

It was Loomis' studies of the human electroencephalogram which earned him a place in the history of neuroscience. He improved on the equipment used by Hans Berger, so that voltage fluctuations on the scalp, reflecting ionic currents in nerve cells (Appendix 5) were recorded with more precision on an eight-foot revolving drum. Although Berger had already noted that brain waves are altered when one briefly goes to sleep, Loomis was able to make all-night recordings. He found a new waveform known as the K-complex, and recognized as well that brainwaves appeared in recurring patterns, thus discovering the principle of sleep having discrete stages (Mendelson, 2017). In 1937 he and colleagues published a paper describing them, which he termed stages A through E. As rapid eye movement sleep was not to be discovered until the 1950s (Chapter

Three), he had in effect defined a first approximation of the stages of what later came to be known as non-REM sleep.

Loomis' interests took him deeper and deeper into physics, and in 1939 he worked with Ernest Lawrence to finance his groundbreaking 184-inch cyclotron. In the late 1930s he was also in touch with European scientists, and the information he learned about the rise of Nazi power led him to believe that war was inevitable. He held the then-unpopular view that America would ultimately become involved, and publicized the need for scientific advances to aid the military. He initially was involved in planning for the Manhattan Project, but Carl Compton, a physicist and president of MIT, convinced him of the importance of also addressing microwave research.

It had been known for some time that radio waves reflecting off objects could in principle be displayed optically, and this seemed to have the potential for building devices to detect ships or planes over long distances. Loomis' interest was heightened in 1940 when Winston Churchill came to New York to show some of the results of British engineering, one of which was the cavity magnetron, which would become the heart of what was later called radar. Loomis' enthusiasm was such that he had outgrown his luxurious 'palace of science' (in Einstein's words) in Tuxedo Park, leading him to establish a new facility in conjunction with MIT in Cambridge. As the war progressed, the 'rad lab' grew to have a staff of almost 4000, including 500 physicists. Their crowning achievement was the development of a practical form of radar, which by the summer of 1942 had greatly decreased the threat of U-boats to Allied shipping, made it possible for pilots to make 'blind landings' in bad weather, and was adapted for automatic control of machine guns in bombers (a project in which neuroscientist Alan Hodgkin was involved;

Chapter Three). He was also instrumental in developing LORAN (originally called 'Loomis Radio Navigation'), a worldwide nautical navigation system which guided Atlantic convoys as well as Navy ships in the Pacific; an improved form operated by the Coast Guard remained operational until it was phased out in favor of satellite navigation in 2010. These inventions had a profound effect on the outcome of the war. A popular saying at the rad lab was that the atomic bomb ended the war, but radar won it.

Figure 1-2: *World War II British naval radar-assisted High Angle Fire Control: The radar operator (below, right) sends information about the range of a radar echo, while the Trainer and Layer (above) send the bearing of a visually seen target, to the gunnery calculating staff. The target's calculated position is then forwarded to the gun crew. The radar display panel (below, left), can provide some information about bearing when the target cannot be seen. While relatively unsophisticated by today's standards, at the time this radar system provided a significant advantage in sea and air warfare.*

After the war, Loomis helped integrate a much smaller rad lab into a peacetime setting; then this man who had never done things half-way turned his attention to his private life. Though always having been seen as outwardly emotionally distant, it turned out that since 1939 he had been having a secret affair with Manette Hobart, the two-decade younger wife of one of his collaborators. He divorced his first wife, remarried, and gave up his luxurious lifestyle and place in society for a quiet, modest life in East Hampton. He avoided publicity and gave no interviews for the next 28 years until his death in 1975.

CHAPTER TWO: THE WARTIME YEARS

William W. Sargant and Eliot Slater, and the treatment of battle fatigue

William Sargant (1907-1988) was born in Highgate, London, into a wealthy business family also noted for its proclivity for producing clergymen. He began his medical education at Cambridge, and later when his family became financially distressed in the 1920s, continued with a rugby scholarship at St. Mary's Hospital (where Alexander Fleming later discovered penicillin). After graduation in 1930 he worked there as a house physician.

In 1934 he was criticized severely for what was seen as his overly aggressive iron treatment for anemia; shortly thereafter he developed what was termed a 'severe mental and physical collapse' and was hospitalized (British Medical Journal, 1988; Dally, 2004). He left St. Mary's and started working at Hanwell Mental Hospital, Middlesex, where he was struck by the lack of efficacy of available treatments, and 'became more and more convinced that insanity would one day appear as a series of physically treatable disorders' (Sargant, 1971). Subsequently he enrolled in a program at the Maudsley Hospital in London, where his keenness for physical and pharmacological treatments in psychiatry was accompanied by a growing disdain for psychoanalysis. He was involved in early studies of amphetamines as a treatment for depression, as well as the use of insulin shock therapy. In 1938 Sargant spent a year on fellowship at Harvard Medical School, and while there met Walter Freeman, an early advocate and practitioner of prefrontal lobotomy, about which he became enthusiastic.

With the onset of World War II, Sargant returned to the Maudsley, which had by this point been separated into two programs, one at Mill Hill School in London, and another in a former workhouse in Sutton, Surrey, which became known as Sutton Emergency Medical Service. Sargent later described the split as partially being a philosophical one; in his words, the 'talkers' went to Mill Hill, while the 'doers' moved to Sutton. Viewing himself much more as a 'doer', he moved to Sutton, whose clinical director was Eliot Slater, with whom he would work closely over the years. At the time, it was managed jointly by the British government for the military patients and the London County Council for civilians. Sargant was often in conflict with the civilian supervising doctors, who were wary of his enthusiastic use of lobotomies and convulsive therapy. After the evacuation of Dunkirk during May and June 1940, the hospital was inundated with new patients, and over the course of the war was responsible for caring for 20,000 casualties (for a discussion of shell shock and battle fatigue please see Appendix 4).

Figure 2-1: British and French ships docking at Dover with evacuees from Dunkirk on May 31, 1940. Approximately 198,000 British and 140,000 French and Belgian soldiers reached England. Facilities such as the Sutton Emergency Medical Service, where William Sargant served, were suddenly faced with large numbers of new patients, many of whom suffered from battle fatigue. Henri Laborit (Chapter Three) was serving on a French destroyer at Dunkirk when it was sunk by torpedoes, and he was rescued by an English sloop.

Eliot Slater (1904-1983) was a person of very different temperament and background, with a father who was an economic historian and a mother who (like Alan Hodgkin's family) was a Quaker and pacifist. He was gentler in his manner, skilled in psychotherapy, and had research interests in the genetics of psychiatric illness. Early in his career in 1934 and 1937 he took leave from the Maudsley to study statistical applications to genetic research in Munich. While there he developed a strong antipathy to the manner in which the Nazis exerted authority over the university. He also met and later married Lydia Pasternak, the sister of Boris Pasternak of later *Doctor Zhivago* fame. On returning to London, he became involved with a Rockefeller Foundation program which brought persecuted Jewish psychiatrists to the Maudsley. Among them was William Mayer-Gross, with whom he later co-edited the well-known textbook *Clinical Psychiatry*, Alfred Meyer, and Eric Guttmann (see later in this Chapter).

When the war broke out, as we mentioned earlier Slater became the medical director of Sutton, as well as St. George's, a major teaching hospital in London. There he found himself responsible for a variety of bright psychiatrists including Denis Full, known for his EEG studies, Kenneth Cameron who became a leader in the field of childhood psychoses, Alexander Kennedy (who later joined the paratroopers to become the first psychiatrist in their long-range missions in Yugoslavia and Greece)—and William Sargant.

At Sutton, Slater provided what Sargant later described as a 'helpful and restraining hand', partially curbing his instincts for rapidly deploying new therapies. With the massive influx of patients in the aftermath of Dunkirk, Sargant was given a much freer hand. He argued that the aggressive manner in which he pushed for physical therapies was partially due to the fact that they were in a psychiatric unit which was part of a general hospital, where he said suicide precautions and similar safety measures were not available. In Sargant's words: 'The war was a great testing time. This was no time to pick and choose our patients; we had to do something, and that our best, for them all' (Sargant and Slater, 1952). He advocated sedation, which he believed might help prevent traumatic memories from becoming entrenched, and used 'sleep treatment', in which intravenous barbiturates were given for prolonged times to Dunkirk casualties.

It was soon recognized that barbiturate-induced sedation would often produce emotional outbursts, in which the patient could be guided by a therapist to re-live the traumatic situation, and afterwards would often appear much less troubled. One British newspaper, which focused on the reliving of the experience rather than the medication, declared: 'Soldiers back from Dunkirk have been cured of nervous disorders by means of hypnotism. Remarkable results were reported this week by Dr William Sargent, a well-known London psychiatric specialist' (Advocate, 1940; see also Bailey, 2014). His advocacy of rapid access to psychiatric care and techniques including sedation were among the influences leading the British to dispatch psychiatrists to the front line areas, first in larger numbers in North Africa and later in Europe. He also developed modified insulin therapy, using lower, sub-coma, doses than previous 'insulin coma' treatment, which was thought to also address the physical aspects of the casualties' condition such as exhaustion and weight loss; the various treatments were often combined (Sargant and Slater, 1952). Electroshock was used for patients with what was then known as endogenous depression, and there was recognition that these various treatments needed to be selected depending on the patient's condition. He believed, for instance, that barbiturate-aided interviews were helpful in an 'hysterical battle casualty' but could be harmful in someone with obsessional illness. Similarly, electroshock was seen as valuable in endogenous depression, but harmful to the stability of the 'constitutional hysteric' (Sargant and Slater, 1952).

By 1944, Sargant had systematized his results, and he and Slater wrote a classic text, *An Introduction to Physical Methods of*

Treatment in Psychiatry, which was translated into a number of other languages, and ultimately came out in five editions until 1972 (Sargant and Slater, 1944). Some years later Sargant recalled Slater apparently calmly writing the chapter on the relation of physical treatments to psychotherapy, while parts of the ceiling were coming down during an attack from German V1 and V2 missiles.

After the war, the Sutton group disbanded. Slater went to the National Hospital for Nervous Diseases, while Sargant became the head of 'psychological medicine' at St. Thomas's Hospital in London, where he remained for the rest of his career. There, freed of Slater's restraining hand and aided by the traditional authority of consultants, he was free to practice what he often described as early intensive treatment which he believed prevented long-term confinement. As time went on many of these such as insulin therapy were discredited or found to be of little benefit. He was also criticized for what to him must have seemed a natural step, evolving his experience in World War II treatments into controversial applications in a new conflict—the Cold War. He was faulted for not having enough concern about obtaining informed consent from patients, and as late as the 1970s, for his disdain for structured clinical trials (Hirsch, 1975).

Slater went on to a distinguished career, in which he was a pioneer in using studies of twins and other approaches to the genetics of psychiatric illness, and was as well the editor of *The British Journal of Psychiatry*. He was also a leader in the movement against capital punishment. In 1977, a few years before his death, he was described as 'England's greatest living psychiatrist'. Though he continued to collaborate with Sargant on new editions of their textbook, and shared a belief in many of the controversial treatments advocated therein, he did not do so with the evangelical zeal and disdain for other viewpoints that were Sargant's style. In short, the two were a study in how very different personalities reacted to the pressures of massive casualties and reduced restraints of wartime. They played a major role in moving Britain away from the then-dominant hold of psychoanalysis, and toward a more biological approach to psychiatry.

German refugee physicians

The growing Nazi control of universities experienced by Eliot Slater while visiting in Munich, and later by Hans Berger in Jena, with accompanying persecution of Jewish psychiatrists and neurologists, led many to emigrate to Britain. As mentioned earlier, this was facilitated by a Rockefeller Foundation program to support their coming to the Maudsley Hospital. The result was a remarkable influx of extremely talented individuals into British medicine. These brief summaries give a sense of their experiences:

Eric Guttmann (1896-1948): After having studied neurology in Berlin as well as psychiatry in Munich under Emil Kraepelin, Eric Guttmann taught in Munich and Breslau until the pressures of the growing Nazi influence led him to move in 1933 to the Maudsley Hospital. There he worked in the Clinical Research Unit until with the outbreak of war the British interned him; he was later freed and worked at the Radcliffe Clinic in Oxford and the Mill Hill Emergency Hospital in London. When the Maudsley Hospital was re-established after the war, he was placed in charge of clinical teaching. He subsequently published with William Sargant on the clinical effects of Benzedrine and played a large role in the acceptance of amphetamines into psychiatry, described 'artificial psychoses' induced by mescaline, characterized the psychiatric consequences of head trauma and the emotional consequences of medical illness. Despite having had to rebuild his life twice--once when emigrating to Britain and again at the outbreak of war when being interned by his adopted country--he achieved more than many who had much longer careers, by the time he perished from heart disease in 1948 at age 52.

Wilhelm (William) Mayer-Gross (1889-1961) was a young physician who took his medical degree in Heidelberg, and began working in the clinic of Franz Nissl, a psychiatrist whose focus was on identifying abnormalities in the tissues of the nervous system in psychoses and dementia. Nissl had gathered about him a coterie of bright young doctors, and was sometimes sad but also sympathetic when some moved into other areas. Among these was Mayer-Gross, who was interested in gathering information on symptoms and patients' expressed thoughts and behavior in order to organize disturbances into syndromes.

When World War I began, Mayer-Gross was sent to the Western Front, and later returned to Heidelberg, where he took care of soldiers who had developed psychiatric disorders. Afterwards he continued to pursue his career in academic psychiatry, and by the early 1930s was chairman of the psychiatry department at the University of Groningen, a public university with roots going back to the seventeenth century. By 1933, persecution by the Nazis had become so difficult that he obtained support from the Rockefeller Foundation to take a position at the Maudsley Hospital in London. There he and Eric Guttmann became well known as teachers and leaders in understanding psychiatric syndromes and developing a methodology for such studies. He also developed a close relationship with Eliot Slater, with whom over the years he published editions of their popular textbook *Clinical Psychiatry.* His interests were varied, and he was also known for his views on a well known psychological analysis of the art of the ancient Minoans.

In 1939 he became research director at the Crichton Royal hospital in Dumfries, Scotland. There he changed his name to William and became a British citizen. Though not having a formal university professorship, he became perhaps the most influential teacher in Britain, combining a German sense of systematization with the British tradition of empiricism (Trail, 1961). In retirement, he took guest professorships at the University of Munich in 1958 and the University of Hamburg in 1960, having come to grips with his own history of mistreatment in the pre-war years. He passed away in 1961, leaving a legacy of having inspired, and provided the structure for continued research, for a generation of younger psychiatrists.

Ludwig Guttmann (1899-1980): The son of an innkeeper and distiller in Upper Silesia (then part of Germany), Guttmann went to medical school, graduating in 1924. A decade later, while he was working as a neurosurgeon at the University of Breslau, he fell victim to the antisemitic Nuremburg Laws, losing his appointment and his standing as a physician, becoming what was known as a '*Krankenbehandler*' (a doctor whose license has been taken away, but was permitted to care for Jews). He became the medical director of the Breslau Jewish Hospital. After *Kristallnacht* on November 9, 1938, in which Nazi SA paramilitaries and civilians raided Jewish homes and businesses, Jews were being rounded up for transport to concentration camps. Guttmann told his doctors to admit anyone who wanted to come into the hospital regardless of their reason. Sixty-four patients were admitted. The following day the Gestapo questioned him about each admission, and only four were sent to the camps; he had managed to save sixty people. A few months later Guttmann was asked to go to Portugal to treat a friend of prime minister Antonio de Oliveira Salazar. On the return trip through London, Guttmann escaped, along with his wife and two children. The family were settled in Oxford by a refugee program. As a refugee, Guttmann was not allowed to practice medicine, so he turned his attention to research, working in the Nuffield Department of Neurosurgery.

At the time, neural injuries were the most disabling form of non-lethal wounds, and at the instigation of the British Army, a special program for studying nerve regeneration was formed at Oxford (Lichtman and Sanes, 2006). It was headed by J.Z. Young, a neurobiologist who was well known for his work on the function of the giant squid axon, used later by Alan Hodgkin and Andrew Huxley in their studies of the propagation of nerve action potentials (Chapter Three). Guttmann contributed to the group's work on documenting the rate of regrowth of damaged neurons, and making the important distinction between growth of the nerve and the recovery of function, which was often some weeks later. The lab greatly contributed to the understanding of how regenerated nerve cells form functioning synapses in order to communicate with muscle cells. They additionally addressed practical issues such as the best sutures and graft materials, which were soon applied in human surgery.

In 1943, the Royal Air Force, concerned about the growing number of injuries to pilots during crash landings, asked Guttmann to lead the first British program devoted to spinal injuries at Stoke Mandeville Hospital. He did so, becoming a British citizen in 1945 and remaining in this post until 1966. In addition to medical treatments, he was known for establishing the Stoke Mandeville Games in 1948, following from his belief that sports were vital to the recovery process of injured soldiers. Within a few years 130 countries were sending paralympic athletes to the annual games, and in 1960 they were first held in parallel to the Summer Olympics in Rome. In 1961 he organized the English Federation of Olympic Sport, and when he later passed away after a heart attack in 1980, had become known as a leader in the founding of sports for persons with disabilities.

Figure 2-2: Russian stamp depicting Ludwig Guttmann as a 'sports legend', after the XI Paralympic Winter Games held in 2014 in the resort city of Sochi.

Eastern European refugees

Ernst Gutmann (1910-1977), a Czechoslovakian physician who had fled to Britain, and was first imprisoned, was assigned to work at the Oxford nerve regeneration laboratory. The staff, who did not know his background, put him to work cleaning the animal facilities. This unusual situation was quickly rectified; he became an integral part of the research team, and was the lead author of one of their first and best known papers (Gutmann et al., 1942).

Frank Berger (1913-2008), also from Czechoslovakia, escaped to Britain in 1939. He is described in more detail in the next chapter because of his post-war discovery of meprobamate, the first modern tranquilizer.

Nathaniel Kleitman (1895-1999) was born in Kishinev, Russia; in his childhood his family fled the pogroms to Palestine. In 1915, early in World War I, he emigrated to the United States, with no funds or support. By 1925 he had a doctorate in physiology and joined the faculty at the University of Chicago. His 1939 book *Sleep and Wakefulness* (Kleitman, 1987 reprint) and his co-discovery of rapid eye movement sleep in 1953 brought about the modern era of sleep research. His work is described in more detail in Mendelson (2017); it should be recognized that he is one more, and very notable, example of the prodigious number of persons displaced by war (in his case, World War I) who became leaders in modern neuroscience.

CHAPTER THREE: THE POST-WAR YEARS

Alan Hodgkin and the neuronal action potential

Alan Hodgkin (1914-1998) was born in Oxfordshire, the son of Quaker parents with pacifist beliefs; during World War I they were subject to a great deal of hostility for their antipathy to the Military Service act of 1916. His father had wanted to go into medicine, but was unable to due to poor vision, and became a banker. He had been very sensitive to suffering, and in 1916 went to Armenia to document the plight of the population at the hands of the Ottomans. He died in June 1918 in Baghdad during a second trip.

Alan's grandfather and uncle had been historians, and he was initially encouraged to follow in their footsteps. He also had an interest in biology, stimulated by an eccentric aunt who taught him to take careful notes while exploring the countryside around her cottage in Northumberland. At age 15 he helped a professional ornithologist study heronries, and later assisted another in studying birds in the marshes of the Norfolk coast. Ultimately, he won a scholarship to Cambridge in zoology, botany and chemistry. Before beginning, he spent a month at a freshwater biological laboratory by Lake Windemere, and two months in Frankfurt with a German family, as it was generally believed that a working knowledge of German was valuable to a career in science. While there he had some exposure to the growing Nazi movement, which influenced his later feelings about war.

As a Cambridge undergraduate he began research projects on frogs' nerves, and as a result was invited to spend a year at the Rockefeller Institute. During that time he also worked at the Woods Hole Marine Biological Laboratory (to which as we described earlier Otto Loewi had become very attached in his later years in America). There he learned about the squid's nerve cell axons, fibers involved in communication between cells, whose remarkably large size greatly facilitated electrophysiologic studies. He went back to Cambridge in 1938, and soon began collaborating with a younger student, Andrew Huxley. They set up shop at the Plymouth Marine Laboratory, where they were able to place a cannula inside squid axons. They then made recordings of the changes in electrical activity across the cell membrane when a neuron becomes excited and sends an electrical impulse down the axon, in the process of communicating with another cell (Appendix 5).

With the outbreak of World War II, Hodgkin enthusiastically joined the effort, his prior experience in Germany having overcome his pacifist upbringing. He first studied issues of oxygenation as well as decompression sickness in flight crews. Later he transferred to the Telecommunications Research Establishment (fitting, perhaps, given his background in cellular communication), and helped design radar systems for night fighter aircraft, and a radar-guided gun turret installed in the tails of bombers, in which he had occasion to fly. In early 1944 he traveled to the U.S. where he visited Alfred Loomis' MIT Radiation Laboratory (Chapter One) to coordinate efforts on radar.

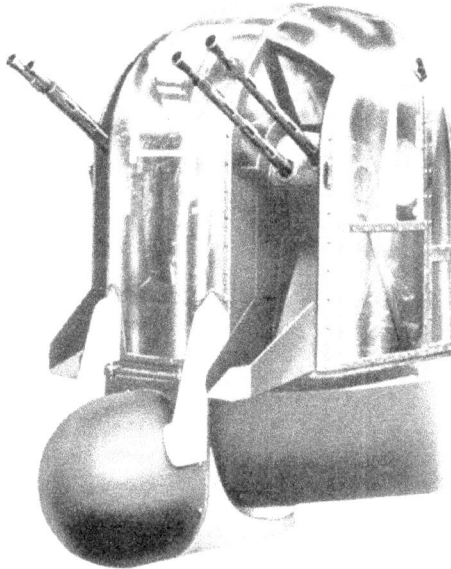

Figure 3-1: Alan Hodgkin contributed to the design of the Automatic Gun-Laying Turret, known as the Village Inn FN121, pictured here in the tail of a Lancaster bomber. In practice, its radar was often used as an early warning system for detecting approaching aircraft rather than for fully automated firing.

Returning to Cambridge after the war, he resumed his work at the marine laboratory with Huxley, and beginning in 1947 they published a series of papers describing the changes in the electrical properties of nerve cells as they communicate with each other (Appendix 5). Their work resulted in the Nobel Prize in Physiology or Medicine in 1963, along with John Eccles who had clarified the role of neurochemicals in transmission of signals between nerve cells. In subsequent years Hodgkin received many honors including becoming president of the Royal Society, and Master of Trinity College, Cambridge. Interestingly, his wartime experiences remained important to him, and apparently were a motivation for writing his autobiography, *Chance and design: Reminiscences of Science in Peace and War* when he was 78 (Hodgkin, 1992). He passed away in Cambridge in 1998 at age 84.

Charles W. Suckling (1920-2013) and halothane

By the 1930s, the limitations of ether and chloroform, discovered almost a century before, were more and more evident, as were those of agents designed to replace them. Ethylene had been developed in 1923 when University of Chicago scientists were trying to determine why carnations were wilting in the university greenhouses ever since gas lighting had been installed (Mendelson, 2020c). It turned out that one of the components of illuminating gas was ethylene, which in addition to being a plant hormone made animals (and later humans) become unconscious. It had many limitations, however: it did not produce adequate muscle relaxation for some forms of surgery, had a significant odor, and though less explosive than ether, was still flammable.

Cyclopropane, developed in the late 1920s and early 1930s, was an improvement in that it had only a mild, non-irritating, petroleum odor, but it was explosive, and additionally some patients developed precipitous drops in blood pressure. By the mid-1930s, the British, with an eye to the likelihood of war, were eager to find a better anesthetic which could be mass-produced. The result was trichloroethylene, which had originally been used as a metal de-greaser and later as a treatment for trigeminal neuralgia before it became an anesthetic (Mendelson, 2020c). It, too, had limitations: it did not evaporate well, was slow in onset, and was less potent in producing deep anesthesia. Moreover, by 1944 it was recognized that some patients receiving it developed nerve palsies, which turned out to be due to toxic decomposition products. There was, then, a great need for a safer and more effective anesthetic. How this came about leads us to the story of Charles W. Suckling.

Born in Middlesex in southeast England in 1920, the son of a clerk in the timber business and later a cargo superintendent, Suckling inherited a family tradition involving the chemistry of explosives. His maternal grandfather and great-uncle had both worked on gun cotton production in Alfred Nobel's factory in Ardeer, Ayrshire and later at the Royal Ordnance factory in Waltham Abbey, Essex, where his paternal grandfather was also employed. Though interested in languages, drama and music, he ultimately enrolled in chemistry at The University of Liverpool in 1939. During the early war years, the Liverpool area was heavily bombed; Suckling's future wife was 'bombed out' twice though not injured. He helped his brother Ted as a fire watcher, and on one memorable occasion the brothers proudly entered a shelter to show off still-hot fragments of an incendiary bomb; he was promptly scolded by his mother, who directed them to take it to the bomb disposal unit outside.

After graduating, the Ministry of Labor and National Service directed Suckling to work at Imperial Chemical Industries, a large company at the time heavily involved in making war materials and pharmaceuticals. As it happens, it had extensive experience with the halogen chemical element fluorine, which was in compounds for aerosols and refrigerants; for this reason, some of its employees, including Suckling, were assigned to the Tube Alloys project, the code name for Britain's secret efforts toward the construction of the atomic bomb. At the time, the emphasis was on a process to enrich the yield of uranium-235. This involved converting 'yellowcake' uranium powder into a gaseous compound containing fluorine, which could then be separated from U-238 in a centrifuge (Figure 3-2). It was in the process of this research that Suckling developed the skills in fluorine chemistry which later served him well in civilian life. In the meantime, in his off-duty hours he served in the Home Guard, and survived the bombing of the Tube Alloys facility at Merseyside, where work was rapidly resumed. It was also in those years that his brother Ted, then in the RAF, perished in a plane crash.

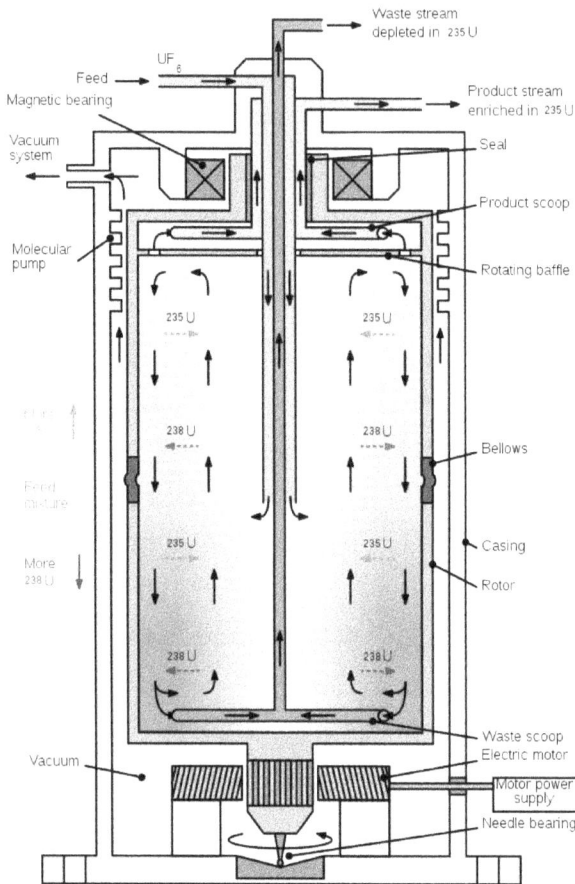

Figure 3-2: Design of gas centrifuge for enriching uranium. Powdered uranium ore ('yellowcake') is first converted to gaseous form by combining it with fluorine, a process studied by Charles Suckling. It is then fed into the centrifuge (top left), in which the U-235 is separated from the heavier U-238, and extracted (top right). The techniques developed in combining fluorine with uranium were later used in making the anesthetic halothane.

After the war, having been in National Service, Suckling was eligible for a grant to return to the University of Liverpool for a PhD. By 1949 he had completed his work and was once again at ICI. There was great interest at the time in finding applications for the new knowledge of fluorine chemistry. Both the management of the Manhattan Project and the Mallinckrodt Company had funded studies of developing anesthetics without favorable results. Nonetheless, one of Suckling's bosses, John Ferguson, was involved in understanding the effects of chemical structure on the potency of compounds producing unconsciousness, and in 1951 stimulated him to see whether halogenated molecules might be developed as improved anesthetics. Suckling was aware that adding halogens to organic compounds reduced their flammability, and that fluorine in particular caused them to vaporize more easily. By 1953, using skills developed during the years in uranium enrichment research, he came up with halothane, which could be considered to be an ethane derivative containing fluorine, bromine and chlorine. It appeared to anesthetize insects and animals, was not explosive, and did not irritate the airways. ICI turned the promising compound over to an anesthesia group at Oxford for further study.

The Oxford group, incidentally, had an interesting history: Lord Nuffield, born William Morris, had been a bicycle and automobile mechanic, who later founded the MG ('Morris Garages') automobile company. This and his other enterprises were very successful, and in 1936 he gave two million pounds to Oxford to establish endowed chairs in four fields of medicine. The field of modern anesthesia was so new that Oxford was reluctant to honor it with a chair, but Lord Nuffield persisted and in 1937 the Nuffield Department of Anesthesia was created. Another one of his creations was the Nuffield Department of Neurosurgery, in which Ludwig Guttmann worked on nerve regeneration (Chapter Two).

By the time the anesthesiologists at Nuffield received halothane, they had established a long history of first testing medicines and procedures on themselves. One of the most colorful examples during the war was Dr. Edgar Pask, who while in the RAF volunteered to be anesthetized until he developed respiratory arrest, in order to test various techniques of artificial respiration. At another time he allowed himself to be anesthetized, then tossed into a London movie studio's water tank, complete with an artificial wave-making apparatus, to determine whether after a plane crash at sea the life jackets could keep an unconscious pilot's head above the water. In this tradition the staff at Nuffield tested halothane on themselves some 40 times before being convinced it was safe enough to give to patients. After a series of clinical studies, they published a paper in the *British Medical Journal*; following this and favorable reports by Michael Johnstone, a Manchester anesthesiologist who was using it in his practice, halothane was soon on the market.

Halothane was not flammable or irritating to the airways and was a precursor to new generations of halogenated anesthetics, though it, too, had limitations, including potential liver damage as well as abnormal heartbeats at high doses (Mendelson, 2020c). It was important in the history of pharmacology, as it was one of the first examples of 'rational pharmacology', that is, constructing a drug based on knowledge of the physiological effects of various chemical structures. The technique with which Charles Suckling designed it was an outgrowth of the development of the atomic bomb. The broader topic of the war effort as a stimulus to other discoveries is considered in Appendix 6.

John Cade and the rediscovery of lithium

John Cade was born in 1912 in Horsham, in western Victoria, Australia. On his father's side was a line of doctors and pharmacists going back almost continuously for 150 years; his mother was a nurse, known for her skill and who had proved her mettle in managing a hospital ward during a typhoid outbreak in 1903. As a child, John was known for his curiosity, his collections of stones and insects, as well as his persistence at whatever endeavor he pursued. When he was three, his father, who had been in the Boer War, and was at that time a 40-year-old general practitioner, volunteered to serve in World War I. He was sent first to Gallipoli, and returned four years later, torn by feelings of needing to make up for lost time but also with fatigue and a sense of failure (de Moore and Westmore, 2016). Suffering from what was known in those years as 'war-weariness', he declined to re-engage in the rigors of private practice, and instead took a series of posts as the doctor at mental hospitals. Living on the hospital grounds, John grew up interacting and playing sports with the patients, and so from an early age became familiar with the consequences of mental illness, and he developed curiosity about its causes.

Following his family tradition, John became a physician, then trained as a psychiatrist. With the outbreak of World War II, he volunteered for the Australian Army Medical Corps, and was sent to Singapore in 1941. With the capture of Singapore by the Japanese in February 1942, he became a prisoner of war at the nearby Changi Prison for three and a half years.

Figure 3-3: Changi Prison when first built in 1936.

Changi prison itself housed 3000 civilians, perhaps five times as many for which it was designed. The Japanese converted the nearby former British Army barracks into a POW camp, which held about 50,000 Allied troops, and interestingly, after the war the British returned the favor, imprisoning Japanese officers there. In these crowded conditions, the malnourished soldiers developed a range of psychiatric symptoms, and Cade, as the only psychiatrist, was put in charge of their care. Certainly, among the conditions he saw were beriberi (from thiamine deficiency), producing difficulty with gait as well as confusion and a type of amnesia, and pellagra (severe niacin deficiency) manifested as skin inflammation, confusion, aggression, and dementia. He was also struck with the fluctuating levels of clarity in some of his other patients, and began to wonder if this might result from varying amounts of some toxic substance, which might accumulate and then be excreted by the body.

After being released in September 1945, Cade was hospitalized to recover from the harsh years at Changi, and then took a post at a veteran's facility, the Bundoora Repatriation Mental Hospital near Melbourne. He pursued his thought that there might be a toxic substance circulating in the blood of psychiatric patients which produced their symptoms, and was particularly interested in the changes in mood characteristic of manic depressive illness (now called bipolar disorder).

Cade set up a laboratory in an unused kitchen, and soon was injecting urine from manic patients into guinea pigs. He found that their urine made the animals much sicker than that from persons without mania, and set about looking for the substance that might be responsible. The most likely candidate was urea, which was the most toxic chemical in urine, but since both patients and non-patients seemed to have equal amounts, he speculated that something else might be making the urea more harmful in the patients. His guess was this might be uric acid, but his efforts to inject it were initially thwarted by its insolubility in water. He found that it could be combined with lithium to make the soluble salt lithium urate. To his surprise, injections of lithium urate did not make the animals sicker, but instead seemed to reduce toxicity, while also making them more quiescent. It turned out that the substance responsible for this quieting effect was the lithium.

Cade took the observation of the guinea pigs quiescence, and translated it into the clinical notion that it might be a useful treatment for mania. After first testing it on himself to establish a dose, he gave it to 10 manic patients, and published the favorable results in an Australian medical journal in 1949. As had happened when similar findings were published by the American physician William A. Hammond in 1871 (Mendelson, 2020a), this was initially met with little interest. Cade went on to fruitless studies of other metallic ions such as rubidium and cerium. He was also discouraged after the death of one patient which was attributed to lithium, and ultimately his interests took him elsewhere.

Lithium had been a component of various consumer products including soft drinks for years. In the weeks preceding the Wall Street crash of 1929, while one company—Alfred Loomis' investment bank-- was quietly liquidating its stocks in favor of gold (Chapter One), another business, The Howdy Corporation, was launching the original 7 Up soft drink. First known as 'Bib-Label Lithiated Lemon-Lime Soda', it contained lithium citrate until it was reformulated in 1948. Lithium chloride was a popular salt substitute, sometimes consumed in large amounts. Toxicity and deaths as a result began to be recognized and in 1949, the same year Cade published his paper, the U.S. Food and Drug Administration removed it from the market.

Thus, it was some years before lithium was pursued further as a potentially helpful medicine. Ultimately a blood test for serum lithium was developed in 1958, making it much safer to use, and other laboratories published larger, well-designed studies. Approximately 20 years after Cade's discovery, it was approved in the U.S. as a treatment for mania, and four years after that for preventive treatment of recurrent episodes. Although a number of other 'mood stabilizer' drugs are now available, it remains a core treatment in the armamentarium for bipolar disorder.

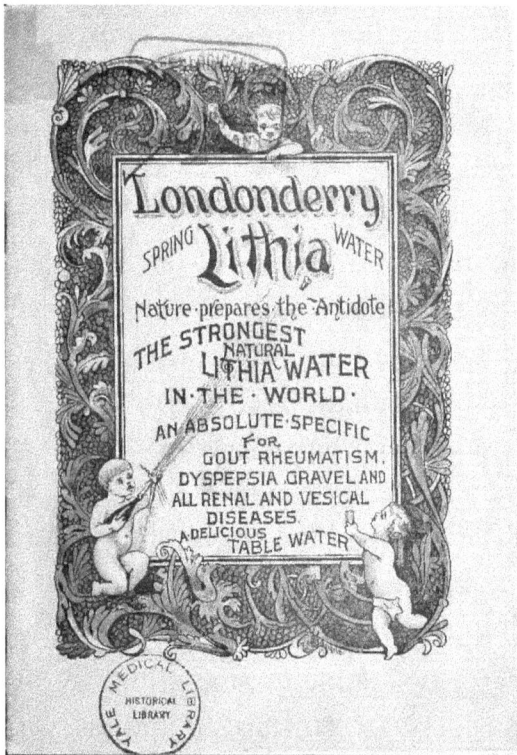

Figure 3-4: Bottled mineral water containing significant amounts of lithium salts was popular in the U.S., particularly from the 1880s until World War I. During the war, new federal regulations began to limit mislabeling and misrepresentation of mineral water products. This and the growing availability of purified tap water led to a decline in consumption. In 1929, lithium reappeared in the earliest version of the soft drink which became 7 Up.

Henri Laborit and 'artificial hibernation'

Henri Laborit (1914-1995) was born in Hanoi in what was then French Indochina, the son of a French army doctor who succumbed to tetanus in 1920. Returning to France, he suffered from tuberculosis and its sequelae, but managed to continue his father's tradition and became a doctor in the French Navy. There he served on the destroyer Siroco, which was sunk by torpedoes during the evacuation of Dunkirk. He was awarded the French Military Cross for his actions during the sinking and subsequent rescue. He spent the remainder of the war in Africa, and the immediate post-war years on a hospital ship near Indochina. He returned, convinced that his chances for advancement would be better as a surgeon, and after training was stationed at the naval hospital near Bizerte, Tunisia.

Figure 3-5: The French destroyer Siroco, launched in 1925, and shown here in 1927. Henri Laborit was aboard when it was sunk by torpedoes on May 31, 1940 during the evacuation of Dunkirk. Over six hundred of the 930 crew and soldiers aboard were lost. Laborit survived after being picked up by an English sloop.

Though Laborit was sometimes in trouble with the military hierarchy—on one occasion the aftermath of a humorous memo put him in hot water—he was very serious about his profession. He came to be very interested in the problem of surgical shock, the often-disastrous drop in blood pressure and circulation which can occur during or after surgery. He developed the notion that one aspect of shock resulted from some of the body's responses to surgery, changes in activity of the autonomic nervous system and the release of the chemical histamine. Among the many actions of histamine is an alteration of the permeability of capillaries and a reduction in blood pressure. He reasoned that if histamine function could be minimized, these effects might not occur. He also recognized that drugs which blocked histamine's action affected the central nervous system and were sedating; thus, they might also make it possible to reduce the doses of anesthetics such as chloroform, which themselves could interact with stress in a way conducive to shock.

Antihistamines had become available in the years following their discovery in 1937 by Daniel Bouvet at Paris's Pasteur Institute. One of the sources was the Rhône-Poulenc company, where the chemist Paul Charpentier was developing a series of them, including promethazine, a phenothiazine derived from the dye methylene blue (Mendelson, 2020b). Laborit found that promethazine given before surgery sedated patients and contributed to the effects of the anesthetics.

In 1950 Laborit was transferred to the Val de Grâce hospital in Paris, which itself had a remarkable history. It was first built as a church and convent in 1645. Over a century later its nuns cared for fallen fighters in the French Revolution, and afterwards it became a military hospital, a tradition which continues to this day. There Laborit continued his interest in antihistamines. He recommended to Rhône-Poulenc that they look further into new compounds which he believed would help stabilize the autonomic nervous system. In response, Paul Charpentier engaged in a project of adding chlorine atoms to antihistamines, a process thought to make them more potent (analogously to Charles Suckling's work of adding halogens to anesthetic molecules, as described earlier in this chapter). By combining chlorine with the antihistamine promazine, he produced a new drug known as chlorpromazine. It had been unsuccessfully tested as a treatment for malaria, but as he found that it had significant antihistaminergic properties, it was sent along to Val de Grâce. Laborit in turn observed that chlorpromazine produced in his patients a state of indifference without necessarily causing them to lose consciousness. Soon it was part of his 'anesthetic cocktail', which lowered the dose of anesthetics needed, and lowered body temperature, a process which Laborit referred to as 'artificial hibernation'.

A common practice in psychiatry at the time was to help control agitated or violent behavior by placing patients in cold water baths, and it occurred to Laborit that chlorpromazine might be useful by producing this same effect pharmacologically. He contacted psychiatric colleagues at Val de Grâce, and soon chlorpromazine was given to Jacques Lh., a 24-year-old patient with a history of mania. His agitated behavior was greatly reduced, and following a three week course of chlorpromazine as well as barbiturates, he was well enough to be discharged. After similar successes with other patients, the results were presented at a medical meeting and published in a 1952 paper, in which chlorpromazine was described as 'a new vegetative stabilizer'; it was met with indifference. There things might have languished, had not an anesthetist colleague mentioned to Laborit that his brother-in-law was Pierre Deniker, a psychiatrist at the prestigious St. Anne's Hospital in Paris. He was contacted, and after Deniker persuaded his boss Jean Delay of its potential, trials with chlorpromazine were begun. It soon became clear that its benefits could be seen without the additional use of barbiturates, and that lowering the body temperature was not necessary for its effect. Most promising, it seemed not only to reduce agitation, but also to help with delusions and aid in disordered thinking.

The enthusiastic reports of Deniker and Delay were echoed by similar findings throughout Europe, and in 1952 chlorpromazine came on the market in France as Largactil ('large in action'). The advent of chlorpromazine, the first modern antipsychotic, ultimately revolutionized psychiatric care. It greatly reduced the use of older physical treatments such as insulin shock therapy and psychosurgery, and made it possible for many hospitalized patients to return to their communities. In 1953, there were approximately 560,000 hospitalized psychiatric patients in the U.S.; by 1975 this declined to 193,000 (Lipton et al., 1978).

Chlorpromazine also led to a change in the way psychiatrists perceived themselves, and the way they were seen by the medical community. Until then, there was a sense of alienation, as psychiatrists worked largely in mental hospitals, distanced from much of the medical world, and relied heavily on a variety of physical treatments. Those that worked in outpatient settings were often practitioners of psychoanalysis, which again seemed far removed from the medical mainstream. Now with the development of chlorpromazine, and the advent of the MAO inhibitor and tricyclic antidepressants in the late 1950s, psychiatrists began to view their work as more like that of other physicians, with an available armamentarium of medicines for treating mental illness.

The discovery of the clinical use of chlorpromazine also coincided with the time of the Korean War, and not surprisingly, Laborit the military physician came up with the idea that it might be useful in dealing with the stress of injury and possibly reducing shock. The U.S. military was receptive to the idea, and soldiers in Korea began to carry chlorpromazine. The result was not as intended—wounded soldiers given the drug became indifferent to their situation and were not always eager to cooperate in rescue. It was suspected that this may have contributed to some deaths, and the project was discontinued (Comité Lyonnais, 2018). Though chlorpromazine continued to revolutionize civilian psychiatric care, in this case a military application was an unfortunate overreach.

Laborit had always been somewhat of a rebel, working mostly outside of major academic institutions. In 1958 he founded his own laboratory in Boucicaut Hospital, largely funded by his patents, and along the way discovered drugs such as the MAO inhibitor minaprine and the sedative clomethiazole. He became a chevalier of the French Legion of Honor in 1953, and in 1957 he won the prestigious Albert Lasker Award along with Pierre Deniker and Heinz Lehmann. He was nominated for the Nobel Prize, but defeated by the longstanding hostility of the Parisian academic community about his views on anesthesia as well as credit for chlorpromazine; indeed, the Dean of the Faculty of Medicine went so far as to travel to Sweden to successfully argue against Laborit receiving it. In 1960 he developed a modern synthesis of gamma-hydroxybutyrate, now used as a treatment for cataplexy and sleepiness in narcolepsy.

Over the years, Laborit's interests expanded to include social issues. He came to believe that an understanding of brain function was essential to dealing with the processes of dominance, aggression and violence. It could be speculated that he was turning the skills and knowledge he had developed over a lifetime to look back on his experiences in wartime as a young man. There was substantial public interest—as well as controversy—and in 1970 he was filmed in an interview with Salvador Dali about his books, and played himself in the 1980 movie *Mon Oncle d'Amerique* with Gérard Depardieu, which highlighted his views. From 1978 to 1983 he elaborated his thoughts as a professor of biopsychosociology at the University of Quebec. He felt to the end, though, that he had not been credited adequately for his scientific work, particularly with chlorpromazine and anesthesia.

Heinz Lehmann and the clinical use of chlorpromazine

Heinz Lehmann (1911-1999) was born in Berlin, the son of a Jewish chest physician and a protestant mother. He had read Freud at the age of 13, and was drawn to psychiatry. In 1937, two years after obtaining his medical degree, the family helped him get Gestapo permission to leave Germany, by obtaining a letter from a friend in Quebec, asking him to come for a ski vacation. He arrived with only his skis and possessions suitable for a two-week vacation. Later his father escaped as well; his mother stayed in Germany, where she died close to the end of the war.

Lehman's refugee status allowed him to obtain only a temporary medical license, and for many years he was required to check in regularly with the Royal Canadian Mounted Police. He chose to work at the Verdun Protestant Hospital (now Douglas hospital), a psychiatric facility for chronic patients, some distance from McGill University, with which it was loosely affiliated. When war came, doctors were scarce, and at one point he was in charge of 600 patients, with the aid of only one nurse and an unprepared staff. Somehow, he managed to engage in a variety of research studies, which he described as being for his own morale. Among these were studies of nitrous oxide for depression, caffeine, and continuous sleep treatment. Until much later in his career, he was isolated from the mainstream of academics in the same way as Henri Laborit, and one may speculate how being on his own in that sense contributed to his methods of thinking (Paris, 1999).

Lehmann kept up with the European literature, and one Sunday in 1952, according to him while reading in a bathtub, he came across one of Deniker and Delay's papers on chlorpromazine. He was quick to obtain it and was struck by its effectiveness, also noting its ability to produce Parkinsonian-like side effects. The first paper describing the studies he conducted with his colleague H.E. Hanrahan came out in 1954, and along the way he invented the word 'antipsychotic' to describe this new class of drugs. In the U.S. there was a fair amount of initial skepticism about chlorpromazine, particularly at universities which continued their focus on both psychoanalysis and behavioral treatments. Much of the enthusiasm came from state hospitals and chronic facilities, who pointed to their ability to discharge chlorpromazine-treated patients who had been hospitalized for months or years. Ultimately the recognition grew, and it came onto the U.S. marketplace in 1955, initially as a drug for nausea and later as an antipsychotic. Lehmann was considered to have played a major role in the acceptance of chlorpromazine in clinical psychiatry, and for this was awarded the Albert Lasker Award, along with Pierre Deniker and Henri Laborit, in 1957. Chlorpromazine continued to transform the practice of psychiatry, in a way as revolutionary as the earlier development of penicillin for the consequences of syphilis on the nervous system (Mendelson, 2020a).

In a manner similar to the story of chlorpromazine, Lehmann also read the European papers about the new tricyclic antidepressant imipramine (Mendelson, 2020a), and again obtained the new drug and published reports leading to its acceptance. He married Annette Joyal, a French-Canadian, and lived on the grounds of the hospital; he was known for his tradition of spending Christmas day going around the wards with his son to visit and try to bring holiday cheer to his patients. A modest man with little interest in physical possessions, he did not have a car, and traveled everywhere by bicycle. He was legally considered a refugee for much of his career, and only obtained a permanent medical license at age 52, and certification by the Royal College of Physicians and Surgeons when he was 60. In older age he was able to once again see Berlin, including the Witte district of his childhood. He continued to take care of patients until a few months before his death at 87.

Frank Berger and meprobamate

Frank Berger (1913-2008) was a young prize-winning physician working as a bacteriologist at the Czechoslovakia National Institute of Health when Hitler's army entered his country on March 15, 1939, about a year after the *Anschluss* which had so affected Otto Loewi and Sigmund Freud in Austria. Berger and his pregnant wife left the same day; denied visas to the U.S., they entered Britain, where for the first few nights he had to sleep on a park bench, while his wife was put up at the Salvation Army. Eventually the Czech Refugee Fund was able to find a hostel for them, and he initially took care of patients at an internment camp. By 1943 he had found work in the British Drug Houses, a conglomeration created in 1908 from drug firms with roots going back to 1714. In 1945 he was working on finding preservatives for penicillin, which was subject to chemical breakdown by bacterial enzymes, when he noticed that rats and mice given one possible preservative became very quiet. When placed on their backs, they would remain in that position, seemingly comfortable, and interestingly without any significant change in their breathing or pulse. Berger, who had always had an interest in anxiety, noted that a similar compound, mephenesin, was already being used for muscle relaxation during surgery, and was known to have what he later called a tranquilizing effect. Mephenesin had limitations as well, however: it was short-acting and was more potent in its effects on muscles than on anxiety. Berger tried to interest his superiors in pursuing a drug which made people more calmer, without avail. Around this time another company developed a better penicillin preservative, and the company moved on to other areas.

In 1947 Berger moved to the U.S., and by 1949 had taken a position as research director at Wallace Laboratories. There he returned to the mephenesin work, and along with a chemist colleague B.J. Ludwig developed meprobamate, which had a longer duration of action and more effects on anxiety. Once again, he was unable to interest his bosses, who were skeptical that there was a market for a medicine whose main purpose was to make people calmer. Undeterred, Berger created a film showing the drug's effects on rhesus monkeys, and showed it as part of a talk at a San Francisco medical meeting. There was great interest, in the face of which the management reconsidered, and by 1955 meprobamate was on the market under the name of Miltown (the area in New Jersey where Berger lived).

Miltown was an immediate success. At one point five percent of Americans were receiving it, and by the end of the 1950s it was the most widely prescribed drug in the U.S. It was talked and joked about on TV, in much the same way as Viagra would be years later. Milton Berle, the comedian whose *Texaco Star Theater* was a hallmark of what has been called the golden age of television, was very enthusiastic about meprobamate; he had previously been known as 'Uncle Miltie', but now facetiously referred to himself as Uncle Miltown.

In time, the drawbacks of meprobamate also became apparent, notably its likelihood of abuse and dependence, as well as the serious consequences of overdose. In 1970 it became a Drug Enforcement Administration Schedule IV agent. Its use gradually declined as doctors began the prescribe benzodiazepines for anxiety, and it remains available, though very infrequently prescribed.

Meprobamate's history is notable for several reasons. It had become the first 'blockbuster' drug in modern psychiatry, in a sense a forerunner of Valium and Prozac. It also influenced the way patients and doctors thought about psychotropic medicines; indeed in 1956 Berger hosted a special meeting of the New York Academy of Sciences with luminaries from many fields including Aldous Huxley, the author of *Brave New World*, to talk about the impact of meprobamate on society. Meprobamate also altered the practice of psychiatry. While chlorpromazine had been oriented to treating hospitalized patients, meprobamate was more applicable to psychiatrists in outpatient clinics. Berger went on to develop the anticonvulsant felbamate, as well as a meprobamate derivative, carisoprodol, prescribed for musculoskeletal pain. He retired from Carter Wallace in 1975, became a professor of psychiatry at Louisville University for some years, and later died after a fall in 2008 at age 94.

Leo Sternbach and the benzodiazepines

Leo Sternbach (1908-2005) was born in an area now in Croatia, to a Polish father and Hungarian mother. As a result of economic difficulties following World War I, the family moved around, ultimately settling in Krakow, where he assisted in his father's pharmacy. After graduating from high school, he was able to enter pharmacy school at the University of Krakow despite its policy of not accepting Jews, based on his father's having practiced as a pharmacist.

Sternbach chose to pursue chemistry as a career, rather than returning to his father's business. After receiving his PhD, he stayed on as a lecturer until 1936, when his position was taken away in favor of a Christian candidate. After finding some temporary positions, he began worsking in 1940 for the pharmaceutical house Hoffmann LaRoche in Basel, Switzerland. With the German invasion of Greece and Yugoslavia in 1941, however, the situation became uncertain, and with the aid of papers prepared by Roche, he fled through France and Portugal, settling at their hew headquarters in Nutley, New Jersey.

In Nutley, Sternbach developed a reputation of being an able chemist, though his advancement was hindered by his proclivity for criticizing his bosses. Fortunately, in those years he made a discovery—a new process for synthesizing biotin (vitamin B7)—which gave him some political protection and a little more freedom to pursue his own interests. With the success of meprobamate in 1955, however, Roche was intrigued with the possibility of finding a new tranquilizer. Several other companies were trying to do this by making small modifications to the meprobamate molecule; Roche asked Sternbach to see if he could find a successor to meprobamate which had an entirely different chemical structure.

Sternbach's first step was to choose a chemical 'building block' for his new project. He settled on a group of azo dyes which he had studied at the University of Krakow (see the companion book *Molecules, Madness, and Malaria,* Mendelson 2020b, for more on the history of the inter-relation of dyes and medicines in psychiatry). At the time he had determined that they were not suitable in practical terms as commercial dyes, published a paper, and moved on to something else. Now he created about 40 variations on those molecules, and had the likeliest ones tested by his behavioral colleagues for tranquilizing properties in animals. They were not successful, and Sternbach's supervisors, feeling that this effort was unproductive, asked him to move on to another project, developing antibiotics. There things might have remained, but in 1957, one of Sternbach's assistants was cleaning out the shelves, and came across an old, unused bottle. It was apparently the 40th, untested dye derivative. Upon being asked what to do with it, Sternbach decided to prevail on his colleagues to do one last behavioral profile. This time it was a success; it produced sedation and muscle relaxation, and unlike chlorpromazine (see section on Henri Laborit) it had minimal effects on the autonomic nervous system. In other animal studies at the San Diego zoo, it reduced aggression in monkeys, transformed a contentious lynx into a frolicsome feline, and had similar effects in baboons, kangaroos and tigers.

The new drug was initially given to humans at too high a dose, producing dizziness and speech difficulties, and a self-administered trial left Sternbach bedridden for two days, much to the alarm of his family. Ultimately the proper doses of this new benzodiazepine drug, chlordiazepoxide, were determined, and it was marketed as Librium in 1960. It was followed a few years later by diazepam (Valium), then in 1970 by flurazepam (Dalmane) for sleep and in 1975 by midazolam (Versed) as an intravenous anesthetic. Eventually more than two dozen related compounds were developed clinically.

The benzodiazepines were perceived to have many advantages over the older barbiturates, and in short order they became the most widely prescribed tranquilizers; from 1969 to 1982 they were the most widely prescribed drugs in America. As time went on, their many limitations began to be recognized. They became Drug Enforcement Administration Schedule IV agents because of the potential for abuse and dependence, and the FDA later posted 'black box' warnings about their many side effects including very toxic interactions with opiates (Mendelson, 2020a). Recognizing these difficulties, physicians continue to shift to other agents such as SSRIs for anxiety and the 'Z drugs' for sleep (Mendelson, 2020a). But for many years Sternbach's decision to test one last drug, when he had been asked to move on to something else, greatly changed the pharmacologic treatment of anxiety.

Sternbach did not rest on his laurels; he went on develop drugs for excessive bleeding during surgery, high blood pressure and other uses, ultimately filing 241 patents. His success with benzodiazepines helped him achieve his ambition, which was to have the freedom to pursue whatever pharmacologic issues interested him. Even after retiring, he maintained an office at Roche, where he worked regularly until about age 95, two years before he passed away.

Arvid Carlsson and the treatment of Parkinson's disease

Arvid Carlsson (1923-2018) grew up in Lund, a small town in southern Sweden, known for its university which goes back to the fifteenth century, at which his father taught history. When he was 16, he and a friend went on a two-week hitchhiking trip in Germany. It was June, 1939, and turned out to be about two and a half months before the war began. In looking back at it in his late seventies, he remembered talking with many Germans, who believed that war would likely start after the summer harvest, and in his words, 'seemed to accept that, though reluctantly' (Carlsson, 2000). On one occasion he stayed in a hostel, and remembered seeing traditionally dressed Jews, studying a book which he took to be the Talmud, in which he guessed they were searching for help in an increasingly perilous time.

After returning to Sweden, Carlsson chose not to follow his two older brothers into his father's profession, and instead enrolled in medical school in Lund in 1941. Though Sweden was neutral, he had a number of experiences with Jewish refugees. In October 1943 the Nazi authorities announced that Jews in occupied Denmark would be deported to Germany in the near future, leading to a large efflux escaping by fishing boat to Sweden, and many settled in Lund. In the spring of 1944, the Swedish royal family got permission from the Germans to send 'white busses' to transport concentration camp inmates to Sweden. About 30,000 were ultimately brought to safety, including 11,000 Jews. A number of them found their way to Lund, where they were housed in tents set up in a park. Carlsson was among the students who examined them, and he vividly described their malnutrition, medical illnesses including tuberculosis, as well as their 'severe anguish and suspiciousness and trusting nobody' (Carlsson, 2000).

Upon earning his MD and PhD in pharmacology, Carlsson stayed on at Lund and later moved in 1959 to the University of Gothenburg. In 1957, dopamine had been detected in the human brain, and shortly thereafter he demonstrated that it functioned as a neurotransmitter.

When brain dopamine was pharmacologically lowered, animals developed symptoms similar to Parkinson's disease, which could be reversed by giving them the dopamine precursor L-DOPA. Before long L-DOPA was being used in humans to treat Parkinson's, and it and subsequent related compounds remain an important part of treatment to this day.

Though there was no obvious connection between Carlsson's wartime experiences and his later discoveries, they clearly affected him deeply. In a way he had some similarities to Alan Hodgkin— both had traveled in Germany in the pre-war era, and remained in their home countries during the war but lived through its consequences. As we will see in the next chapter, in both cases they returned to their experiences when in older age.

Figure 3-6: Campus of the University of Lund, 2005. Founded in 1666 and with roots going back to 1425, it is one of the oldest universities in northern Europe.

Because of its location near the Öresund, the narrow strait separating Sweden from Denmark, Lund received a number of refugees during the war. Many were well educated and took positions teaching at the university. Arvid Carlsson, while a medical student there, examined rescued concentration camp prisoners.

CHAPTER FOUR: SUMMARY

Effects of the war years on the lives of the discoverers

The experiences of the future leaders in neuroscience and psychopharmacology were remarkably variable. Often the result was displacement, as in the young medical graduate Heinz Lehmann's emigration to Canada under the pretext of going on a skiing vacation. Frank Berger was a young physician who had recently won a national prize for developing antisera for salmonella when he was forced to flee the German occupation of Czechoslovakia in 1939 and ended up sleeping on park benches in London. Well-established psychiatrists such as William Mayer-Gross and Eric Guttmann moved to England to avoid the growing Nazi influence on their hospitals and universities.

Often the effects of hostilities were much more devastating: Otto Loewi was imprisoned, lost all his financial assets and even when released and emigrating to England and the US, was separated from his wife until 1941. Hans Berger, discoverer of the EEG, became enmeshed with the SS activities at his university, ultimately became depressed and hung himself in his own clinic. Some put themselves at risk in an effort to help others: Ludwig Guttmann and Viktor Frankl (later in this chapter) brought danger upon themselves by admitting Jewish patients to their hospitals and altering diagnoses as a way of saving them from euthanasia or being sent to the camps.

For others, the advent of war led to the willing application of their skills to the effort: John Cade, a psychiatrist and the son of a physician who had served in World War I, had been in the Australian Militia since 1935, and in 1940 volunteered for the Australian Army; after his capture in Singapore in 1942 he continued to work as psychiatrist for his fellow POWs in a Japanese prison. Alan Hodgkin, who had seen Nazi repression firsthand during a stay in Germany in 1932 joined the Royal Aircraft Establishment, where he worked on issues of oxygen supply for pilots, and later, radar. Alfred Loomis, who had worked in ballistics in World War I and among other things classified sleep stages in the 1930s, turned his efforts to radar and marine navigation systems.

An Oxford laboratory studying nerve regeneration (Chapter Three) exemplified the ironies that were often seen in those years: it included Ludwig Guttman, a German refugee neurosurgeon mentioned above, as well as the Czech physician Ernst Gutmann, a POW of the British who was originally recruited to the project to help clean the animal facilities, as well as Peter Madawar, who later went on to win the 1960 Nobel Prize for studies of transplantation immunity. The first World War also had its share of ironies: Robert Barany, a civilian surgeon in the Austrian army, received word in 1914 that he had won the Nobel Prize for his work on the vestibular apparatus while in a Russian POW camp. His wartime experiences led to new discoveries as well, as he treated neurological injuries that led to better understanding of the vestibular system and the cerebellum.

When looking at the lives of these discoverers, it's clear that they were a very diverse group, coming from a variety of backgrounds. Many did not steer a straight course in coming to medicine or science. Otto Loewi, like neuroanatomist Santiago Ramon y Cajal before him (Mendelson, 2018), was primarily interested in art, while Hans Berger was intrigued by astronomy. Alan Hodgkin came from a family of historians and it was thought that he would continue in this tradition. Sigmund Freud planned to study law. Charles Suckling was interested in languages, drama and music. One thing they all had in common, though, was that once in their careers they displayed remarkable resilience when faced with adversity. As we described earlier, many had suffered persecution, jail or displacement. One, Eric Guttmann, was interned once again in his new country when war broke out; others like Heinz Lehmann were given limited rights as refugees for many years. Ludwig Guttmann as a refugee was not allowed to practice medicine. Others such as Freud were old and ill when fleeing to Britain. John Cade returned in poor health after three and a half years in a prisoner of war camp. Yet with few exceptions, they were able to move beyond what had happened to them and lead remarkably productive lives.

The experience of the war years seemed to stay with them, though, and sometimes emerged in older age. Alan Hodgkin dealt with his memories of those years in his last book, *Chance and Design: Reminiscences of Science in Peace and War,* written when he was 78. Arvid Carlsson described at length his youthful travel to Germany and his impressions of the escaped concentration camp prisoners in his Nobel Prize biography at age 77. Similarly, Henri Laborit wrote extensively about applying the principles of neuroscience to issues of aggression and violence beginning in his 60s. All three had moved beyond the war to outstanding careers, but they appeared to carry inside them the memories of those years, and revisited them later in life.

Viktor Frankl

Viktor Frankl, psychiatrist and Holocaust survivor, is not included in the main body of this book, which is devoted to neuroscience and psychopharmacology. But no account of the war years on foundational psychiatrists would be complete without mentioning his experiences. He was born in 1905 in Prague, the son of an official in the Ministry of Social Service. During the first World War the family fell on hard times, and the children were sent to local farms to beg for food. After the war, during high school, he became interested in the mind and began corresponding with Sigmund Freud. By 15 he had given a public lecture, 'On the meaning of life'. He later had a falling out with Freud, becoming interested in the Adlerian approach involving the inferiority complex as well as the importance of the role of the community, but later moved away from Adler as well because of his emphasis on the need for meaning as a crucial drive. He completed his medical degree in 1930; by 1937 he was practicing neurology and psychiatry in Vienna, only to see his work stopped by the Nazis after the *Anschluss*. By 1940 he was the director of neurology at the Rothschild Hospital, where in a manner reminiscent of Ludwig Guttmann at the Breslau Jewish Hospital he attempted to prevent Nazi euthanasia of the mentally ill by altering the diagnoses. He was given a visa to the U.S., but did not use it as he did not wish to leave his aging parents.

In 1942 Frankl, his wife, and parents were arrested by the Gestapo and sent to the camps, where, like John Cade as a POW, he organized efforts to deal with the psychological consequences of internment, and recorded his experiences. After liberation in 1945 he became the medical director of a hospital for displaced persons. In his search for his family, from whom he had become separated, he learned that his wife, brother and mother had been killed in Auschwitz. In 1946 he assembled many of his thoughts during an intense nine-day period resulting in *Man's Search for Meaning*. In it he emphasized that the pursuit of meaning was central to those who survived, and that meaning could be found in love, constructive work, and courage during hardship. It came out in English in 1959 and in later editions (Frankl, 2006) and was widely read. By the 1990s a Library of Congress survey noted it as one of the 10 most influential books in the U.S. (Fein, 1991).

Warfare in human experience

Warfare can be viewed in many different ways, and some have emphasized how deeply it is embedded in human experience. In a recent book, Margaret MacMillan concluded that 'War is not an aberration, best forgotten as quickly as possible. Nor is it simply an absence of peace which is really the normal state of affairs' (MacMillan, 2020). It has also been thought of as a 'man-made public health problem' (Razum et al., 2019), a view which suggests it can be approached or mitigated in the same way, for instance, as infectious diseases. Thoughts about it and what it might mean go back to the Iliad in the 8th century BC, and undoubtedly will continue for a long time to come. World War II in particular has been proclaimed to have its own unique features, as the last global conflict between nation-states with 'mega-machine' social organization. It has been viewed by some as a 'scientists' war', in which the development of penicillin, radar and the atomic bomb played major roles, as did the widespread use of psychoactive substances such as amphetamines (Rasmussen, 2011). As a matter of timing, it is evident that modern neuroscience and psychopharmacology greatly moved forward in the years after the war. Many of those who made the discoveries were affected by the conflict in a variety of ways, often with displacement, persecution and suffering, sometimes tragedy, and occasionally with new insights. This book has been written so that as we debate the broader issues, sometimes in fairly abstract terms, we do not lose sight of the personal experiences of the people involved.

Related topics and suggested readings

In closing, it's worthwhile to mention that are many related topics about pharmacology and warfare. They are beyond the scope of this book, which is devoted to stories of the lives of the discoverers, but it is worthwhile mentioning them and providing references as starting points for those who wish to read further.

The historical use of drugs to enhance soldierly skills: The Viking 'berserkers' beginning in the ninth century were reputed to fight in a trance-like state and to be impervious to pain, perhaps with the aid of hallucinogenic mushrooms. In World War II, the German army overran France partially with the aid of Pervitin, a methamphetamine-based drug which aided in staying awake; the widespread use of methamphetamine in *Fliegerschokolade* ('flyer's chocolate') and *Panzerschokolade* ('tanker's chocolate') were well known. Some authors have emphasized the widespread use of Pervitin and the opiate Eukodol among the Nazi leaders (Ohler, 2018).

Though stimulants had been used by Allied aircrews on long missions, in the Vietnam War the military provided substantial amounts of stimulants, steroids and neuroleptics in what has been called by some the first 'pharmacological war' (Kamienski, 2016). In Chapter Three we mentioned Henri Laborit's idea that chlorpromazine might be of aid in dealing with the stress of combat; in this case the notion was less than successful, as wounded soldiers in the Korean War, when given chlorpromazine, became so indifferent after receiving the medication that they did not participate in their own rescues, and the program was discontinued. The use of drugs to improve soldiers' performance is an important topic, and one which needs to be confronted, but it is beyond the scope of this book, which is devoted to the World War II experiences of the people who went on to make basic discoveries in neuroscience and pharmacology. It is mentioned only for the purpose of perspective and to point to literature on related subjects.

Psychochemical warfare, the use of non-lethal psychotropic compounds to incapacitate the enemy is again outside the scope of this book, but is a notable related topic. Its history goes back to at least 184 BC, when Hannibal's army deployed belladonna as a weapon (Fitzgerald, 2018), an application which was explored in more modern militaries as late as in twentieth century America. In the early seventeenth century, Powhatan the chief of Algonquin-speaking tribes in what is now Virginia and the father of Pocahontas, is said to have weaponized hallucinogenic plants in his struggles with the Jamestown colonists (Price, 2003).

Moving ahead three centuries, in the 1950s, amphetamines, cocaine and nicotine were considered by the U.S. military for airborne deployment but found to lack sufficient potency; barbiturates and opiates were also rejected, and although psychedelic agents such as LSD were explored in the secret CIA MKUltra program for purposes of interrogation, it was felt that their effects were too unpredictable for the battlefield (Fitzgerald, 2018). The agent felt to have promise was a refined version of the belladonna which had been familiar to Hannibal—the anticholinergic agent 3-quinuclidinyl benzilate, which was stockpiled by the U.S. and incorporated into the M43 BZ cluster bomb and the M44 generator cluster in the early 1960s. Effects were thought to appear in 3-6 hours after exposure, and to last up to five days (Goodman, 2010). They also had many limitations including the visibility of the deployed cloud, and the effectiveness of simple defensive measures such as breathing through several layers of cloth. Neither bomb was viewed as an integral aspect of U.S. chemical weaponry, and they were destroyed in 1989. Thus was the short modern military career of a substance which blocked the '*Vagusstoff* ' of Otto Loewi's dreams (Chapter One).

CLUSTER, BOMB: INCAPACITATING, BZ, 750-POUND, M43

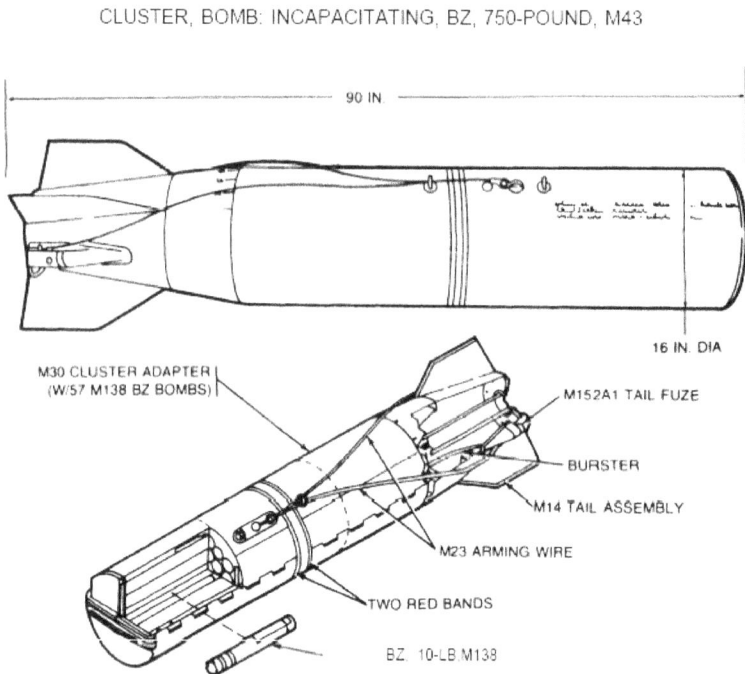

90 IN.

16 IN. DIA

M30 CLUSTER ADAPTER
(W/57 M138 BZ BOMBS)

M152A1 TAIL FUZE

BURSTER

M14 TAIL ASSEMBLY

M23 ARMING WIRE

TWO RED BANDS

BZ. 10-LB.M138

Figure 4-1: The M43 BZ cluster bomb, produced beginning in 1962, held three stacks of 19 bomblets, each of which contained six oz. of 3-quinuclidinyl benzilate. It was estimated that the anticholinergic effects including confusion and disorientation could incapacitate 94 percent of enemy troops in an area up to about two acres. It was categorized as an 'interim weapon', and was never deployed.

Hitler's vendetta against the Nobel Prize: This apparently began after his experience with Carl von Ossietzky, who was offered the Nobel Peace Prize of 1935 for articles which raised awareness of Germany's secret rearmament. Ossietzky was forbidden to go to Norway to accept it. Debilitated by years in prison camps, and hospitalized for tuberculosis, he issued a statement denouncing the Nazi policy that acceptance of the prize would place a recipient outside German society. His health never recovered, and he died in a Berlin hospital in 1938 while still under police surveillance. We mentioned this in the story of Otto Loewi, who was freed from jail and allowed to emigrate only after turning over the money from his Nobel Prize to a Nazi-controlled bank, but there were a number of other cases as well. Among these was Gerhard Domagk, who developed the first sulfa antibacterials. When offered the Prize in Physiology or Medicine in 1939, he was briefly jailed and forced to sign a statement renouncing it, and was only able to accept it in 1947 (Mendelson, 2020b). More information about the relation of German scientists to the Nazi regime and Hitler's repression of winners of the Nobel Prize is described in articles devoted to the topic (Crawford, 2000).

APPENDIX

1. A brief history of neuroscience and psychopharmacology through the post-World War II years

The brain is in a very fundamental way unlike organs such as the lungs or heart. In the latter it is possible to intuit aspects of their function just by looking at them. Centuries ago, for instance, anatomists could speculate that the lungs worked something like a bellows and that the heart resembles a pump. In contrast, the brain appears to be an aggregate of jelly-like substance to which historically it was difficult to recognize function from its outward form. By the 1830s, advances in microscopy made it possible for Matthias Jakob Schleiden and Theodor Schwann to determine that cells were the basic units of structure in plants and animals, but it was less clear that this applied to the brain. Josef von Gerlach in 1871 famously proposed a 'reticular' theory, that nerve cells were all connected together into one complex network. This issue persisted when Camillo Golgi, the inventor of the silver stain which greatly advanced visualization of neurons, still believed in their continuity, while his co-Nobel Prize winner in 1906, Santiago Ramon y Cajal, argued that they were discrete entities, connected to others by synapses, specialized structures for communication.

Charles Sherrington and Edgar Adrian clarified further the integration of the nervous system and the function of synapses. In the 1920s Adrian demonstrated the 'all or none' response of a stimulated neuron. There was an ongoing controversy as to whether the communication between cells was primarily electrical or chemical, and in 1921 Otto Loewi carried out his famous experiment in frogs' hearts indicating that a chemical substance, later identified as acetylcholine, was released by the vagus nerve in transmitting its signal to heart tissue (Chapter One).

The late 1920s also saw the discovery of the human EEG by Hans Berger, though it was not well recognized and used internationally for another decade. In 1937 Alfred Loomis and colleagues used the EEG to make the first systematic description of sleep stages (Chapter One). In 1942, J.Z. Young's group at Oxford described the process of nerve cell regeneration and the establishment of functioning synapses between regenerated cells and muscle tissue (Chapter Two). William Sargant and Eliot Slater clarified the need for rapid treatment of soldiers suffering shell shock and developed the barbiturate abreaction treatment technique.

The aftermath of World War II saw the flowering of both basic neuroscience and psychopharmacology (Chapter Three). Alan Hodgkin and Andrew Huxley resumed work they had begun in 1939 on signaling in the giant squid axon, and ultimately described the flow of ions across membranes during the action potential. In 1949 John Cade wrote his first paper on lithium as a treatment for bipolar disorder, and 1952 Henri Laborit's initial studies of chlorpromazine ultimately led to its entering the market as the first modern antipsychotic. In 1951, John Carew Eccles provided further evidence that most nerve cells communicated by chemical means, and a variety of studies demonstrated the importance of neurotransmitters. Among them, in 1957 Julius Axelrod showed that drugs which prevented the action of the enzyme monoamine oxidase led to increased quantities of catecholamines in the synaptic space, a cornerstone of the 'monoamine hypothesis' that amounts of related chemicals were important in mood regulation (Mendelson, 2020a). By the late 1950s Nathan Kline's work with iproniazid and Roland Kuhn's studies of imipramine led to the clinical release of the first MAO inhibitor and tricyclic antidepressant respectively. The appearance of meprobamate in 1955 and chlordiazepoxide in 1960 ushered in the beginnings of the modern pharmacology of anxiety.

The 1950s also marked the discovery of human rapid eye movement sleep by Nathaniel Kleitman and Eugene Aserinsky. Returning to the basic sciences, Arvid Carlsson, working with an animal model of Parkinson's disease, showed that L-DOPA could reduce symptoms, and James Austin reported that steroids reduced inflammation in neurons. Bengt Falck and Nils-Ake Hillarp reported in 1962 that when nerve tissue is exposed to formaldehyde, neurons employing biogenic amine neurotransmitters would fluoresce, making it possible to trace their pathways in the brain. This was a key step in moving beyond the notion that sleep is a passive state resulting from lack of sensory stimulation to a newer view that it is a physiological process actively regulated by neural pathways containing biogenic amines and other neurotransmitters (Mendelson, 2017). Michael Kidd and Robert Terry used the electron microscope to demonstrate the plaques and neurofibrillary tangles seen in Alzheimer's disease.

All of this activity and more in the post-World War II years led to the creation of neuroscience and psychopharmacology as scientific disciplines. In the early 1950s, psychiatry, at least in the U.S., was dominated by Freudian psychodynamic thought (Appendix 3). Although chlorpromazine and the new antidepressants were initially viewed with skepticism, their remarkable effectiveness ultimately resulted in newfound respect, and researchers began to think of themselves of psychopharmacologists. Though studies of the nervous system were somewhat overshadowed by physics as a prominent discipline in the immediate post-war years, the popularity of neuroscience as a research endeavor and the development of training programs rapidly grew. The two fields saw the formation of their own professional organizations, the American College of Neuropsychopharmacology in 1961 and the Society for Neuroscience in 1969.

2. Narrative and neuroscience

Narrative—the presentation of a series of related experiences— has long had a place as one of the 'rhetorical modes', that is, ways in which language-based communication takes place. In the early 19th century, it was formulated as one of the four basic methods, along with argumentation, exposition and description (Connors, 1981). One example of the utility of narrative is the Bible, which in addition to its importance as the bedrock of Christianity is a masterpiece of literature. By one estimate, nearly four billion copies have been sold globally in the last five decades (Polland, 2012). Much of its content is presented in the form of stories and parables; one wonders whether the kind of enduring readership it has had across two millennia could have been maintained if instead its ideas were expressed purely in the form of a religious philosophical tract.

In more modern times it has been suggested that there is something about narrative that is a particularly good fit for the human nervous system as a way of organizing information, for instance, in considering one's past, thinking about the future, or trying to grasp someone else's perspective. There is increasing evidence that brain cortical regions which are activated when comprehending narratives overlap with those of the 'default mode network' which is evident when someone's attention is not centered on the environment (Buckner et al., 2008; Yuan et al., 2018). Among these are the medial temporal lobe, involved in memories and associations with past events, the medial prefrontal area, which may have a role in flexibly applying this information to mental constructions, and integrative areas such as the posterior cingulate gyrus (Buckner et al., 2008). This has led to speculation that in the resting state, the process of the mind wandering involves an activity similar to building narratives (Jacobs and Willems, 2018). Some neuroscientists have gone so far as to suggest that they should use stories to improve communication between science and society (Martinez-Conde et al., 2019). There is growing recognition, then, that when communicating with general audiences, the emphasis should be placed less on presenting the greatest amount of data, and more on how it is presented. For those who would like to read more narratives of the lives of pivotal figures in neuroscience, many who were alive in the 1990s provided their own memories in the much more detailed, six-volume collection *The History of Neuroscience in Autobiography* (Society for Neuroscience, 2021).

3. Effects of World War I on Freud's view of psychoanalysis

The World War I years had a profound effect on the directions of Freud's thinking in at least two ways. In his earlier years, he had been heavily influenced by two very different views in eighteenth and nineteenth century German thinking. On the one hand, idealists such as Immanuel Kant and G.W.F. Hegel pictured the mind as completely transparent and available to a person; on the other hand, some such as Arthur Schopenhauer and Friedrich Nietzsche believed that we are driven by feelings and urges of which we are unaware, and over which we have little control. Freud's thought in the area of 1905 was part of the movement away from idealism (Shaw, 2021). He viewed the mind in what has been called a topographical model, which he likened to an iceberg: on the surface is the conscious part, while beneath the waterline rests the unconscious, where shameful, largely sexual impulses are held at bay with the help of the preconscious, a kind of gate which regulates what materials are allowed to ascend to consciousness (Figure A-1). Neuroses were viewed as the price being paid to keep these seemingly unacceptable thoughts under control.

By the end of the war, Freud turned his attention to the importance of a destructive force as well, and emphasized an ongoing struggle between 'eros', the drive toward love and life-sustaining activities such as eating and sex, and 'thanatos', the drive for violence and death. When thanatos is directed toward others, it is expressed as aggression; when turned inward it results in self-destruction. He went on to evolve other ways of picturing the mind, notably in the 1920s as having a tripartite structure including the id, which houses the instincts, the ego, which grounds a person in reality, and the superego, the source of morality. From that point on, though, he believed that an understanding of humans must include recognition of an inherent predilection for aggression.

Conscious

Preconscious   EGO

SUPEREGO

ID

Unconscious

Figure A-1: Visual representation of some of the features of the mind as described by Sigmund Freud. The topographical model, showing the unconscious, preconscious, and conscious areas is represented with the image of an iceberg. As Freud's thinking evolved, he added a more structural view with the 'psychic apparatus', made up of the ego, superego, and id.

The second influence of the World War I years on Freud's thinking is the recognition of man as a social animal. His earlier focus had been on individuals, particularly their childhood development and relation to their parents, and there had been more of a willingness to see these processes as being basically biological in origin. After the war Freud began to recognize the importance of how societies are organized and affect individual behavior. His twin 1915 essays

Thoughts for the times on war and death expressed a progressively darker view of humanity and disappointment in the inability of society to provide adequate safeguards. He continued to try and relate how society helps shape the personality, culminating in his 1930 work *Civilization and its discontents*. In it he described social influences on the development of the superego, which contains accepted ideals and can also be the source of feelings of guilt.

4. Shell shock and battle fatigue in World Wars I and II

The history of psychiatric responses to trauma go back to mankind's earliest recorded experiences. The Sumerian epic of Gilgamesh, dating from the second millennium BC, centers on the reaction of the hero to the death of his companion Enkidu, resulting first in grief, but then in depressed mood, recurring dreams of the gods condemning Enkidu to death, and a fixation on his own mortality (Figure A-2). Descriptions of soldiers whose emotional response to battle led to their being sent home are found in Deuteronomy 20: 1-9. The 5th century BC Greek historian Herodotus describes combatants who suddenly became blind, though they had not received any physical injury (Crocq and Crocq, 2000).

Figure A-2: Gilgamesh (bearded figure on the right, between two lion-headed figures), as seen in a Mesopotamian seal impression. It is probable that he was a historical king in Sumer, located in southern Mesopotamia (now Iraq), whose story became mythologized after his death.

For the purposes of the time scale in this book we will begin with World War I, which has sometimes been referred to as the first war in which killing was carried out on an industrial scale. On both sides it soon became clear that many soldiers suffered from an acute reaction which could involve blindness, loss of speech or paralysis, but also anxiety, loss of appetite, heart palpitations, insomnia and nightmares. It could be found in up to 40 percent of casualties, and at the battle of the Somme, for instance, may have affected 16,000 or more soldiers (BBC, 2004).

Initially, shell shock, as it was called, was considered to literally be due to neurologic complications from the physical shock of explosions, but later it was recognized to appear in those who had not been near detonations. The prevailing view was that it was due to a weakness or cowardice, and dealt with by trials for desertion, and sometimes execution. Others viewed it as a form of hysteria to which persons with 'degenerate' nervous systems were predisposed (Bradley, 2014). Even as it came to be viewed as a condition resulting from the extreme stress of warfare, which required treatment, methods often included shaming and solitary confinement. Charles Myers, a British physician in the Royal Army Medical Corps, who popularized the term shell shock in an article in Lancet in 1915, later thought it might not be accurate, and recommended hypnosis. Physical treatments for what became known as 'war neurosis' were also developed. Lewis Yealland, a Canadian-born physician working in London, advocated electrical means. If a soldier appeared unable to speak, for instance, electrodes were applied to the throat, or if he were paralyzed, they were placed on the limb (Bailey, 2014). In some cases, the extremes of treatment were later recognized. Julius Wagner-Jauregg, an Austrian psychiatrist who later won the Nobel Prize for malaria therapy of neurosyphilis and later yet became a Nazi supporter, was charged after the war with having used techniques that approached torture, including an early form of electroshock that sometimes was lethal. As described earlier, among those who testified as an expert witness for the state was Sigmund Freud.

In the years between the wars, the British military emphasized improving screening and higher standards for recruitment rather than treatment later. At the beginning of World War II there were about six psychiatrists in the military, but as the need became clear this rose by 1944 to about 200, half of whom were dispatched overseas (Bailey, 2014). As described in Chapter Two, the evacuation of Dunkirk in May-June 1940 produced a large influx of patients with what became known as 'battle fatigue' or 'combat stress reaction'. Among the many places they were sent was the Sutton Emergency Hospital in Surrey, where William Sargant and others found the need to treat large numbers quickly. Sargant's belief was that it was important to prevent traumatic memories from becoming entrenched, and he employed sedation in an effort to interrupt that process. Often the sedated soldiers would have strong emotional reactions involving their memories, and a technique of 'abreaction' was developed in which a therapist would guide them through recalling their experience, as a method of catharsis. A modified form of insulin therapy was developed, in which sub-coma doses were used in an attempt to reverse physical effects of deprivation and inanition during combat. Other facilities such as Hollymoor Hospital in Birmingham used physical techniques, but also began to explore the application of group therapy. The wards were seen as therapeutic communities in which patients would be supportive of each other. Sargant and others also emphasized treatment as close to the front lines as possible, and as the war went on, psychiatrists were dispatched to North Africa and later to Europe. The principle came to be known as PIES, referring to proximity, immediacy, expectancy, and simplicity. More recent experiences in forward psychiatry are described in *Psychiatrists in Combat* (Ritchie, Warner, and McLay, 2017).

The evolution of understanding psychiatric war casualties since World War II is beyond the scope of this book, but is considered in a number of sources (Friedman, 2021). The prominence of seemingly neurologic symptoms such as blindness or deafness during World War I had already declined and now continued to do so, to the degree that they had at most a very minor role, and ultimately disappeared as a requirement for the diagnosis. It also became clear that this disorder was not limited to the military, but was common in the civilian population. The American Psychiatric Association's first compendium of psychiatric disorders, the Diagnostic and Statistical Manual in 1952, referred to gross stress reaction, and by the third edition in 1980 described it as post-traumatic stress disorder, or PTSD. By the fifth edition (American Psychiatric Association, 2013) it no longer appeared as a type of anxiety disorder; recognizing that it is a complex state including changes in mood, cognition, and arousal, as well as anger and impulsive behavior, it was placed in the category of 'Trauma and Stressor-Related Behaviors'. This represented a movement toward emphasizing the role of the traumatic event which is outside the individual, rather than a weakness in the person involved.

The diagnosis of PTSD involves the presence of a number of features; among the most prominent ones are having experienced a catastrophic event, intrusive recollections which may be in the form of nightmares or flashbacks, avoidance of situations which are reminders of the trauma, changes in mood, heightened arousal resulting in hypervigilance, and other symptoms. In order to be considered PTSD, these symptoms must have lasted at least one month, cause distress by reduced functioning in one's social or occupational life, and not be explainable as the effects of medicines, abused substances or other illnesses (Friedman, 2021).

Longitudinal studies have found that PTSD as it is described now can be a chronic condition which lasts for years. It is accompanied by changes in the nervous system, including hyperarousal of the sympathetic system, alterations in central endocrine processes, and overactivity of brain structures such as the amygdala (Shiromani et al., 2009). It may affect about seven percent of the population at some time in their lives (Gradus, 2021).

5. Hodgkin and Huxley's studies of nerve cell action potentials

After the war, Alan Hodgkin and Andrew Huxley re-started their collaboration on recordings of electrical activity in the giant squid axon. In a series of papers beginning in 1949 they reported that when a nerve fiber is conducting an impulse, it does so by a very brief (perhaps one-thousandth of a second) reversal of the difference in electrical charge between the outside and inside of a nerve cell, producing a wave of excitation traveling the length of the fiber. They demonstrated that the pattern of this 'action potential' results from altered permeability to sodium, potassium and chloride ions. It turned out that during the beginning of depolarization (lowering of the electrical difference between the outside and inside of the cell), there is an increase in flow of sodium ions into the nerve cell, which depolarizes the membrane even further, a process later referred to as the 'Hodgkin cycle'. After the action potential, an active protein transport system returns sodium and potassium ions to their resting concentrations. Hodgkin and Huxley also speculated about the presence of ion channels on the surface of nerve cells, which were confirmed some years later.

6. The war effort as a stimulus for discovery

Just as the wartime experiences of the discoverers were variable, so was the degree to which the needs of war stimulated neuroscience and pharmacology. As we have seen, the initial discoveries of chlorpromazine and lithium were done in the setting of military hospitals, though by rather independent individuals and not necessarily as part of an organized war effort. This happened in other areas of pharmacology as well, for instance when studies of mustard gas as a chemical weapon led to the development of the first cancer chemotherapeutic drugs, the nitrogen mustards (Wilke, 2019; Sloan Kettering Institute, 2021). Again, this had not been the intent of the wartime research, which was devoted to weapon-building and defense, and came about because of a connection made by creative individuals.

In other cases, the discovery of new drugs was very much by intent: the preparation for impending hostilities led to the creation of trichloroethylene, as a less flammable anesthetic which might have fewer of the drawbacks of cyclopropane and chloroform. Alternatively, sometimes studies as part of the war effort led to technical advances that later were applied to civilian uses: thus, knowledge gained from halogenating uranium compounds to enrich

U-235 for atomic weapons was subsequently utilized by Charles Suckling to make the anesthetic halothane (Chapter Three). Finally, of course, the urgent needs of dealing with casualties led directly to advances in treatments in neurology and psychiatry. Thus, the wartime studies by J.Z. Young's group at Oxford led to both a better understanding of the principles of nerve regeneration and practical applications for treating the wounded. As mentioned earlier, Peter Madawar, one of the Oxford group, also worked on another project of wartime importance—skin grafting—which led to his later studies of transplant immunology and winning the Nobel Prize.

Sometimes, the results of working with casualties were mixed: the massive influx of soldiers with battle fatigue after Dunkirk led William Sargant to develop barbiturate abreaction, modified insulin therapy and sleep therapy, though arguably in an atmosphere in which the lack of safety precautions and informed consent would make the modern reader uncomfortable. Many of these approaches have now been discarded, as were the much more drastic treatments for shell shock utilized by Julius Wagner-Jauregg in World War I. Other principles which were developed, such as the benefits of rapid treatment close to the battlefield, have proved to have more lasting value.

In more recent years, some have sounded alarms over what they see as the 'militarization of neuroscience', pointing to contemporary studies of controlling weapons by human thought, or using the MRI for 'brain fingerprinting' to search for terrorists at airports (Gusterson, 2007). Others have argued that the roots of neuroscience have always been both military and civilian, pointing to the role of facilities such as the Walter Reed Army Institute of Research and concluding that 'Neuroscience was funded and shaped to meet the needs of warfare and national security imperatives' (Howell, 2020).

The developments in neuroscience in the three-quarters of a century since 1945 are well beyond the scope of this book. Looking at the experiences of the people involved during the World War II era, however, one would have to say that all in all, the roots were more complex. Summarizing what we have described above, while some discoveries were made by organized wartime programs, others were the unplanned results of technologies developed for other purposes during the war. In still other cases military hospitals provided the settings in which individuals pursued their goals, which were not specifically for military applications. Sometimes the clinical experience of wartime medicine influenced the thinking which led to later discoveries, or to post-war projects such as the Paralympic Games. Often the need to do military service put off research which could not be continued until after the war, and sometimes a repressive regime prevented acknowledgement of new discoveries. There was no single pathway to the growth of neuroscience and psychopharmacology during the wartime era. In a sense it has the feeling of an organic process, influenced in many different ways by its environment.

It's also interesting to consider what paths neuroscience and psychiatry might have taken in the absence of the world wars. As we've talked about earlier, the World War I experience profoundly affected Freud's conception of the mind, turning away from a more biologically influenced topography and emphasizing the role of an aggressive, violent drive and the relation of the individual to society. One wonders in what directions his thoughts might have taken him had he not had this experience. Might psychoanalytical theory have developed, for instance, in a direction more compatible with his original roots in neurology? And though we have talked about possible ways war may have stimulated new findings, one could speculate on how many innovations were not made because their potential discoverers did not survive.

REFERENCES

Note regarding referencing: In some cases, online materials do not provide a specific date of the original posting. In these situations, the year listed is the date the site was accessed.

Advocate newspaper, Burnie, Tas., Sept. 2, 1940: Dunkirk soldiers cured by hypnotism.
https://trove.nla.gov.au/newspaper/article/68377829

American Psychiatric Association: Diagnostic and Statistical Manual, 5th edition. American Psychiatric Publishing, 2013.
https://www.amazon.com/Diagnostic-Statistical-Manual-Mental-Disorders/dp/0890425558/ref=sr_1_1?crid=5T6VQ42LQEJE&dchild=1&keywords=dsm+5+diagnostic+and+statistical+manual+of+mental+disorders&qid=1610209723&s=books&sprefix=DSM%2Caps%2C231&sr=1-1

Bailey, R.: The second world war: shellshock to psychiatry. Transcript of a lecture for Gresham College at the Museum of London, March 17, 2014.
https://www.gresham.ac.uk/lecture/transcript/download/the-second-world-war-shellshock-to-psychiatry/ (Accessed January 8, 2021)

BBC: Shell Shock. March 3, 2004.
http://www.bbc.co.uk/insideout/extra/series-1/shell_shocked.shtml (Accessed January 8, 2021)

British Medical Journal: Obituary: William W. Sargant. 1988:297:789. https://www.bmj.com/content/297/6651/789

Bradley, J.: Shell shock treatments reveal the conflict in psychiatry's heart. The Conversation, August 5, 2014. https://theconversation.com/shell-shock-treatments-reveal-the-conflict-in-psychiatrys-heart-29822 Accessed (Accessed January 8, 2021)

Braslow, J.T. and Marder, S.R.: History of psychopharmacology. Ann. Rev. Clin. Psychol. 15: 25-50, 2019. https://pubmed.ncbi.nlm.nih.gov/30786241/

Buckner, R.L. et al.: The brain's default network: anatomy, function, and relevance to disease. Ann. N.Y. Acad. Sci. 1124: 1-38, 2008. https://pubmed.ncbi.nlm.nih.gov/18400922/

Butler, A.: *The Lives of the Saints: Complete Edition.* Catholic Way Publishing, 2015. https://www.amazon.com/Lives-Saints-Reverend-Alban-Butler-ebook/dp/B00UXKORWE

Carlsson, A.: Arvid Carlsson – Biographical. NobelPrize.org. (Written at the time of the award in 2000). Online citation: Nobel Media AB 2021. Sun. 24 Jan 2021. https://www.nobelprize.org/prizes/medicine/2000/carlsson/biographical/

Cohen, D.: *Escape of Sigmund Freud.* Abrams Press, 2012. https://www.amazon.com/Escape-Sigmund-Freud-David-Cohen/dp/1590206738

Comité Lyonnais Recherches et Thérapeutiques en Psychiatrie: *The birth of psychpharmacotherapy: Explorations in a new world, 1952-1968*, in Healy, D.: *The Psychopharmacologists*. Vol. 3, CRC Press, 2018, pp 1-54.

Connors, R.J.: The rise and fall of the modes of discourse. College Composition and Communication. 32: 444-455, 1981.
https://www.jstor.org/stable/356607?origin=crossref&seq=1

Crawford, E.: German scientists and Hitler's vendetta against the Nobel Prizes. Historical Studies in the Physical and Biological Sciences 31: 37-53, 2000.
https://doi.org/10.2307/27757845

Crocq, M-A and Crocq, L.: From shell shock and war neurosis to posttraumatic stress disorder. Dialogues in Clinical Neuroscience 2: 47-55, 2000.
https://www.researchgate.net/publication/51751154_From_shell_shock_and_war_neurosis_to_posttraumatic_stress_disorder_a_history_of_Psychotraumatology (Accessed January 8, 2021)

Dally, A.: Sargant, William Walters. Oxford Dictionary of National Biography, 2004.
https://www.oxforddnb.com/view/10.1093/ref:odnb/9780198614128.001.0001/odnb-9780198614128-e-40195

Danto, E.A.: Trauma and the state with Sigmund Freud as Witness. Int. J. Law Psychiatry 48: 50-56, 2016.
https://pubmed.ncbi.nlm.nih.gov/27324417/

De Moore, G. and Westmore, A.: *Finding sanity: John Cade, lithium and the taming of bipolar disorder.* Allen & Unwin, 2016.

Fein, E.B.: Book Notes. The New York Times, Nov. 20, 1991. https://www.nytimes.com/1991/11/20/books/book-notes-059091.html (Accessed January 23, 2021)

Fitzgerald, G.M.: Agent 15 poisoning. Medscape, Dec. 18, 2018. https://emedicine.medscape.com/article/833238-overview

Frankl, V.: *Man's Search for Meaning.* Beacon Press, 2006. https://www.amazon.com/Mans-Search-Meaning-Viktor-Frankl/dp/0807014273/ref=sr_1_1?crid=JUH9PF12ZCG1&dchild=1&keywords=man%27s+search+for+meaning&qid=1611165658&s=books&sprefix=man%27s+search%2Cstripbooks%2C206&sr=1-1

Freud, S.: *Thoughts for the Times on War and Death.* https://www.loc.gov/exhibits/freud/ex/175.html

Freud Library 12: Freud, S.: *Civilization Society and Religion.* Penguin, 1985. https://www.amazon.com/Library-Civilization-Society-Religion-Pelican/dp/0140217452

Friedman, M.J.: PTSD history and overview. U.S. Dept. of Veterans Affairs, National Center for PTSD. https://www.ptsd.va.gov/professional/treat/essentials/history_ptsd.asp (Accessed January 8, 2021)

Gay, P. (ed), Freud, S. (author): *The Freud Reader.* W.W. Norton, 1995.

https://www.amazon.com/Freud-Reader-Sigmund/dp/0393314030/ref=sr_1_1?crid=WK07T6ZQO6UE&dchild=1&keywords=project+for+a+scientific+psychology+freud&qid=1611247644&sprefix=project+for+a+scientifi%2Caps%2C208&sr=8-1

Glickstein, M.: *Neuroscience, a Historical Introduction.* MIT Press, 2014.
https://mitpress.mit.edu/books/neuroscience

Goodman E, Ketchum J, Kirby R (ed.). Historical Contributions to the Human Toxicology of Atropine. Eximdyne. 2010.
https://www.amazon.com/Historical-Contributions-Toxicology-Atropine-2010-05-19/dp/B01K8ZFBYQ/ref=sr_1_3?dchild=1&keywords=eximdyne&qid=1612370977&s=books&sr=1-3

Gradus, J.L.: Epidemiology of PTSD. U.S. Department of Veterans Affairs, National Center for PTSD.
https://www.ptsd.va.gov/professional/treat/essentials/epidemiology.asp#backtotop (Accessed January 9, 2021)

Gusterson, H.: The militarization of neuroscience. Bulletin of the Atomic Scientists, April 9, 2007.
https://thebulletin.org/2007/04/the-militarization-of-neuroscience/

Gutmann, E. et al.: The rate of regeneration of nerve. J. Exp. Biol. 19: 14-44, 1942.
https://jeb.biologists.org/content/209/18/3485

Hassabis, D. et al.: Imagine all the people: How the brain creates and uses personality models to predict behavior. *Cereb Cortex* 24:1979–1987, 2014.
https://pubmed.ncbi.nlm.nih.gov/23463340/

Healy, D.: *The Psychopharmacologists.* Vol. 3, CRC Press, 2018, pp 1-54.

Hirsch S. (1979) Psychiatric 'Experiments'—Moral and Ethical Issues. In: Current Themes in Psychiatry 2. Palgrave Macmillan, London.
https://doi.org/10.1007/978-1-349-04494-8_8

Hodgkin, A.: *Chance and Design: Reminiscences of Science in Peace and War.* Cambridge University Press, 1992.

Howell, A. 'Neuroscience hasn't been weaponized – it's been a tool of war from the start.' The Conversation, 2020.
https://theconversation.com/neuroscience-hasnt-been-weaponized-its-been-a-tool-of-war-from-the-start-69097

Jacobs, A.M and Willems, R.M. The fictive brain: neurocognitive correlates of engagement in literature. Rev Gen Psychol 22:147–160, 2018.
https://journals.sagepub.com/doi/10.1037/gpr0000106

Kamienski, L.: The drugs that built a super soldier. The Atlantic, April 8, 2016.
https://www.theatlantic.com/health/archive/2016/04/the-drugs-that-built-a-super-soldier/477183/

Kleitman, N.: *Sleep and wakefulness*, University of Chicago Press, reprinted 1987.
https://www.amazon.com/Wakefulness-Midway-Reprint-Nathaniel-Kleitman-dp-0226440737/dp/0226440737/ref=mt_other?_encoding=UTF8&me=&qid=1611954204#reader_0226440737

Latson, J.: Why Freud chose Nazi Germany over America. Time, May 6, 2015.
https://time.com/3840374/freud-birthday-anniversary-history/

Lev, E.: Jewish Medical Practitioners in the Medieval Muslim World: a Collective Biography. Edinburgh University Press, April 30, 2021.
https://www.amazon.com/Jewish-Medical-Practitioners-Medieval-Muslim/dp/1474483976

Lichtman, J.W. and Sanes, J.R.: J. Experimental Biol. 209: 3485-3487, 2006.
https://jeb.biologists.org/content/209/18/3485

Lipton, M. et al.: in *Psyhopharmacology: A generation of progress*. Raven Press, New York, 1978, p. 4.

Macmillan, M.: *War: How Conflict Shaped Us*. Random House, 2020.
https://www.amazon.com/War-How-Conflict-Shaped-Us/dp/1984856138/ref=sr_1_1?crid=1RRH7UTLEV3ME&dchild=1&keywords=war+how+conflict+shaped+us&qid=1610818699&s=books&sprefix=war%3A+how+%2Caps%2C207&sr=1-1

Martinez-Conde, S. et al.: The Storytelling brain: how neuroscience stories help bridge the gap between research and society. J. Neurosci. 39:8285-8290, 2019. DOI:
https://doi.org/10.1523/JNEUROSCI.1180-19.2019

Mendelson, W.B.: *The Science of Sleep,* University of Chicago Press, Chicago, 2017.
https://www.amazon.com/Science-Sleep-What-Works-Matters/dp/022638716X/ref=sr_1_3?crid=3DI3SSR3H9YZG&dchild=1&keywords=the+science+of+sleep&qid=1611683209&s=books&sprefix=the+science+of+sleep%2Caps%2C205&sr=1-3

Mendelson, W.B.: *Understanding Antidepressants,* 2018.
https://www.amazon.com/Understanding-Antidepressants-Wallace-B-Mendelson-ebook/dp/B07B4GWKSN/ref=sr_1_1?dchild=1&keywords=Understanding+antidepressants&qid=1611676456&s=books&sr=1-1

Mendelson, W.B.: *The Curious History of Medicines in Psychiatry.* Pythagoras Press, 2020a.
https://www.amazon.com/Curious-History-Medicines-Psychiatry-ebook/dp/B083ZRMCW1/ref=sr_1_1?dchild=1&keywords=The+curious+history+of+medicines+in+psychiatry&qid=1609261419&s=digital-text&sr=1-1

Mendelson, W.B.: *Molecules, Madness, and Malaria: How Victorian Fabric Dyes Evolved into Modern Medicines for Mental Illness and Infectious Disease.* Pythagoras Press, 2020b.
https://www.amazon.com/Molecules-Madness-Malaria-Victorian-infectious-

ebook/dp/B088QPG14K/ref=sr_1_1?dchild=1&keywords=Molecules%2C+Madness%2C+and+Malaria&qid=1610732386&s=books&sr=1-1

Mendelson, W.B.: *Nepenthe's Children: The History of the Discoveries of Medicines for Sleep and Anesthesia.* Pythagoras Press, 2020c.
https://www.amazon.com/Nepenthes-Children-discoveries-medicines-anesthesia-ebook/dp/B08H4XDZBN/ref=sr_1_1?dchild=1&keywords=Nepenthe%27s+children&qid=1610732459&s=digital-text&sr=1-1

Ninivaggi, F.J.: Why War? Psychology Today, May 26, 2012.
https://www.psychologytoday.com/us/blog/envy/201205/why-war

Ohler, N.: *Blitzed: Drugs in the Third Reich*, Mariner Books, 2018

Paris, J.: Heinz Lehmann: A pioneer in modern psychiatry. Canadian J. Psychiat. 44: 441-442, 1999.
https://journals.sagepub.com/doi/pdf/10.1177/070674379904400503 (Accessed January 9, 2021)

Polland, J.: The 10 most read books in the world. Business Insider, Dec. 27, 2012.
https://www.businessinsider.com/the-top-10-most-read-books-in-the-world-infographic-2012-12

Price, David A.: *Love and Hate in Jamestown: John Smith, Pocahontas, and the Heart of a New Nation,* New York: Knopf, 2003, pg .204.

Randall, A.: *Black Bottom Saints.* Amistad, 2020.

http://www.amazon.com/Alice-Randall/e/B001IXS1SQ%3Fref=dbs_a_mng_rwt_scns_share

Rasmussen, N.: Medical science and the military: The Allies' use of amphetamine in World War II. J. Interdisciplinary History 42: 205-233, 2011.
https://www.jstor.org/stable/41291190

Razum, O. et al.: Is war a man-made public health problem? The Lancet 394:1613. November 2, 2019. DOI:
https://doi.org/10.1016/S0140-6736(19)31900-2

Ritchie, E.C., Warner, C.H., and McLay, R.N.: *Psychiatrists in Combat: Mental Health Clinicians' Experiences in the War Zone.* Springer, 2017.
https://www.amazon.com/Psychiatrists-Combat-Mental-Clinicians-Experiences-ebook/dp/B072DRHPL1

Sargant, W. and Slater, E.: The influence of the 1939-1945 war on British psychiatry. *Comptes Rendus des Seances, Premier Congres Mondial de Psychiatrie,* 1950. Hermann & Cie, Editeurs, Paris, 1952.

Sargant, W.: *The Unquiet Mind.,* Macmillan, 1971.

Shaw, B.: Historical context for the writings of Sigmund Freud. Columbia College: The Core Curriculum.
https://www.college.columbia.edu/core/content/writings-sigmund-freud/context (Accessed January 9, 2021)

Shiromani, P.J. et al. (eds.): *Post-traumatic stress disorder: Basic science and clinical practice.* Humana Press, New York, 2009.

Shoup, J.R.: *A Collective biography of Twelve World-Class Leaders.* University Press of America, 2005.
https://www.amazon.com/Collective-Biography-Twelve-World-Class-Leaders/dp/0761831592

Sloan Kettering Institute: 1945-1959: Birth of Chemotherapy.
https://www.mskcc.org/research/ski/about/story-ski (Accessed February 3, 2021)

Society for Neuroscience: The History of Neuroscience in Autobiography. L.R. Squire, Ed.
https://www.sfn.org/About/History-of-Neuroscience/Autobiographical-Chapters (Accessed January 16, 2021)

Trail, R.R.: William Mayer-Gross. Royal College of Physicians. Vol. V, p 275, 1961.
https://history.rcplondon.ac.uk/inspiring-physicians/william-mayer-gross

United States Holocaust Memorial Museum, Washington, D.C.: Sigmund Freud. Accessed January 25, 2021.
https://encyclopedia.ushmm.org/content/en/article/sigmund-freud

Tucker, N.: Freud's last days in Vienna as Nazis approached. Library of Congress Blog, September 23, 2019.
https://blogs.loc.gov/loc/2019/09/freuds-last-days-in-vienna-as-nazis-approached/

Wilke, C: From Chemical weapon to chemotherapy, 1917-1946. The Scientist, April 1, 2019.

https://www.the-scientist.com/foundations/from-chemical-weapon-to-chemotherapy--19171946-65655 (Accessed January 9, 2019)

Yuan, Y. et al.: Storytelling is intrinsically mentalistic: A functional magnetic resonance imaging study of narrative production across modalities. J Cogn Neurosci 30:1298–1314, 2018.
https://pubmed.ncbi.nlm.nih.gov/29916789/

Zeidman, L.A.; Stone, J.; Kondziella, D.: New Revelations About Hans Berger, Father of the Electroencephalogram (EEG), and His Ties to the Third Reich. Journal of Child Neurology, Volume: 29 issue: 7, page(s): 1002-1010.
https://doi.org/10.1177/0883073813486558

PICTURE CREDITS

The author's assessment is that all images are in the public domain or presented under the terms of Section 107 of the U.S. Copyright Law (the 'Fair Use' provision). When appropriate, all reasonable efforts have been employed to trace copyright holders and to get their permission for the use of copyright material. The author apologizes for any errors or omissions in this list and will gratefully include any corrections in future editions if notified.

Cover: 'View along the River Thames in London toward smoke rising from the London docks after an air raid during the Blitz', September 7, 1940. National Archives and Records Administration and New York Times Paris Bureau Collection, from Wikimedia Commons, which states 'This work is in the public domain in the United States because it is a work prepared by an officer or employee of the United States Government as part of that person's official duties under the terms of *Title 17, Chapter 1, Section 105 of the US Code.*'
https://commons.wikimedia.org/wiki/File:London_Blitz_791940.jpg

Figure Pre-1: (Paracelsus) Paracelsus, Der ander Theyl der grossen. Credit: Wellcome Collection. Attribution 4.0 International (CC BY 4.0)

Figure P-1: (Sigmund and Anna Freud). Unknown author, from Wikimedia Commons, which states 'This work is in the public domain in its country of origin and other countries and areas where the copyright term is the author's life plus 70 years or fewer.'

Figure P-2: (Freud's couch) Robert Huffstutter, from Wikimedia Commons, which states 'This file is licensed under the Creative Commons Attribution 2.0 Generic license.'

Figure 1-1: (Loewi's Nobel Prize) Wellcome Collection. Creative Commons Attribution 4.0 International CC BY 4.0

Figure 1-2: (Naval radar): Royal Navy Official Photographer, in Wikimedia Commons, which states 'This work created by the United Kingdom Government is in the public domain.'

Figure 2-1: (Dunkirk evacuation): War Office official photographer, in Wikimedia Commons, which states 'This work created by the United Kingdom Government is in the public domain.'

Figure 2-2: (Ludwig Guttmann stamp): Unknown author, from Wikimedia Commons, which states 'This work is not an object of copyright according to article 1259 of Book IV of the Civil Code of the Russian Federation No. 230-FZ of December 18, 2006.'

Figure 3-1: (Village Inn turret). British Government, from Wikimedia Commons, which states 'This work created by the United Kingdom Government is in the public domain.'

Figure 3-2: (Gas centrifuge): Inductiveload, from Wikimedia Commons, which states 'I, the copyright holder of this work, release this work into the public domain. This applies worldwide.'

Figure 3-3: (Ghangi prison): The Straits Times, from Wikimedia Commons, which states 'This work formerly enjoyed copyright in Singapore but is now in the public domain because its term of copyright has expired.'

Figure 3-4: (Lithia water, 1891): Londonderry Lithia Spring Water Co., from Wikimedia Commons, which states 'This work is in the public domain in the United States because it was published (or registered with the U.S. Copyright Office) before January 1, 1926.'

Figure 3-5: (Siroco, 1927): U.S. Navy, from Wikimedia Commons, which states 'This file is a work of a sailor or employee of the U.S. Navy, taken or made as part of that person's official duties. As a work of the U.S. federal government, it is in the public domain in the United States.'

Figure 3-6: (University of Lund): Mucklpu, from Wikimedia Commons, which states 'This file is licensed under the Creative Commons Attribution-Share Alike 3.0 Unported license.'

Figure 4-1: (M43 BZ cluster bomb, 1982): U.S. Army, from Wikimedia Commons, which states 'This image is a work of a U.S. Army soldier or employee, taken or made as part of that person's official duties. As a work of the U.S. federal government, the image is in the public domain.'

Figure A-1: (Iceberg): Historicair, from Wikimedia Commons, which states 'I, the copyright holder of this work, release this work into the public domain. This applies worldwide.'

Figure **A-2:** (Gilgamesh): Dr. L. Legrain, in Wikimedia Commons, which states 'This file is made available under the Creative Commons CC0 1.0 Universal Public Domain Dedication.'

www.ingramcontent.com/pod-product-compliance
Lightning Source LLC
Chambersburg PA
CBHW060619210326
41520CB00010B/1394